Diversity in STEM

Diversity in STEM: Analyzing Inequities and Improving Opportunities in Education and the Workplace offers a survey of diversity in the broad field of Science, Technology, Engineering, and Mathematics (STEM) and provides potential solutions to improve outcomes in education, industry, and society. Offering a U.S.-based point of view, but with globally applicable concepts around race, gender, culture, politics, and socioeconomics, the book identifies where issues around diversity in STEM exist, how they were created, and how these issues are being addressed in STEM education and the STEM workforce.

Features:

- Identifies conditions and causes of inequities from a societal perspective.
- Offers guidelines and solutions to identify and address cultural gaps in STEM.
- Covers STEM initiatives implemented at the K-12, college, and vocational levels and how they are beginning to alter the STEM landscape.
- Illustrates the benefits of fostering and maintaining a diverse, equitable, and inclusive workforce.
- Explores best practices used by companies and organizations to recruit, support, and develop diverse talent and strategies to continually evolve.
- Guides and empowers STEM professionals to seek out organizations whose values are aligned with their own.

Providing an analytical and constructively practical viewpoint, the authors offer readers across the sciences, engineering, and medicine, as well as policymakers, the opportunity to consider why diversity and equity in STEM matter and how to apply best practices that support inclusivity to ensure successful outcomes for individuals, organizations, and society.

Benjamin Blocker II has worked for over 25 years in information technology for corporate organizations, as well as an entrepreneur and instructor. Dr. Blocker's experience includes multimedia development, development of information systems, enterprise resource planning, and eLearning systems development. He also has experience with business analysis and business development, as well as biomedical research relating to human physiology, agricultural science, and animal bioscience. His experience has involved clinical research as well as the development and implementation of related information systems and software. Dr. Blocker holds a doctorate in information systems and communications from Robert Morris University, as well as a master of software engineering, master of science in information science, and a bachelor of science in agricultural sciences, all from Pennsylvania State University.

What Every Engineer Should Know

Series Editor: *Phillip A. Laplante, Pennsylvania State University*

For more information about this series, please visit: www.routledge.com/What-Every-Engineer-Should-Know/book-series/CRCWEESK

Diversity in STEM
Analyzing Inequities and Improving Opportunities in Education and the Workplace

Benjamin Blocker II

CRC Press
Taylor & Francis Group
Boca Raton London New York

CRC Press is an imprint of the
Taylor & Francis Group, an **informa** business

Designed cover image: © 2026 Shutterstock

First edition published 2026
by CRC Press
2385 NW Executive Center Drive, Suite 320, Boca Raton FL 33431

and by CRC Press
4 Park Square, Milton Park, Abingdon, Oxon, OX14 4RN

CRC Press is an imprint of Taylor & Francis Group, LLC

© 2026 Benjamin Blocker II

Library of Congress Cataloging-in-Publication Data
Names: Blocker, Benjamin, author.
Title: Diversity in STEM : analyzing inequities and improving opportunities in education and the workplace / Benjamin Blocker.
Other titles: Diversity in science, technology, engineering, and mathematics
Description: First edition. | Boca Raton : CRC Press, 2025.
Identifiers: LCCN 2024051006 (print) | LCCN 2024051007 (ebook) | ISBN 9781032592138 (hbk) | ISBN 9781032592152 (pbk) | ISBN 9781003453581 (ebk)
Subjects: LCSH: Women in science. | Minorities in science. | Women in technology. | Minorities in technology. | Diversity in the workplace.
Classification: LCC Q130 .B585 2025 (print) | LCC Q130 (ebook) | DDC 331.08—dc23/eng/20250123
LC record available at https://lccn.loc.gov/2024051006
LC ebook record available at https://lccn.loc.gov/2024051007

ISBN: 978-1-032-59213-8 (hbk)
ISBN: 978-1-032-59215-2 (pbk)
ISBN: 978-1-003-45358-1 (ebk)

DOI: 10.1201/978-1-003-45358-1

Typeset in Times
by Apex CoVantage, LLC

Contents

Series Statement

What Every Engineer Should Know includes an overwhelming catalogue of information. Regardless of discipline, engineering intersects most scientific fields and modern technologies. Practicing engineers, however, must also navigate managerial, socio-economic, and even political concerns. The engineer discovers soon after graduation that any curriculum omits important and thorny issues of daily practice— for example, problems concerning new technologies, scientific advances, business practices, legal implications, and team dynamics.

With the *What Every Engineer Should Know* series of concise, easy-to-understand volumes, every engineer can access primers on important subjects across a broad range of knowledge areas, including intellectual property, contracts, software, business communication, management science, and risk analysis, as well as more specific topics such as embedded systems design. These books are very accessible to every engineer, scientist, and technology professional and are necessary to remain competitive in this dynamic, global economy.

Preface

In a world that is as fast evolving as ours, Science, Technology, Engineering, and Mathematics (STEM) are the four central pillars. Advancements in medicine, renewable energy, artificial intelligence, and the list goes on and on demonstrate the amazing abilities of STEM. But one question has hovered overhead. The matter of diversity, or more specifically, the absence of it. This book hopes to highlight the tangled and knotted history of U.S. diversity issues in STEM, illuminating not just where these occurrences are but how they came to be and continue to exist.

To start, we will look at the conditions and root causes of the diversity dilemma in STEM that got us here. We will explore race, gender, culture, and politics and socioeconomics. These dimensions interplay in intricate ways to create obstacles that prevent equal access to STEM careers and studies. By looking at historical context and modern-day analysis we will see how these barriers were built and the systemic forces that are still active today.

Historically, race and ethnicity have played a significant role in creating disparities in access to STEM opportunities. Decades of rooted historical biases and systemic racism have tilted the playing field against people of color, leading to stark underrepresentation in STEM jobs. Adding yet another layer of complexity is gender dynamics. Despite recent improvements in gender equity, women, particularly women of color, are acutely underrepresented in some STEM areas. Added to that, cultural expectations and stereotypes have long perpetuated such inequities, and any act to step out of expected norms has so far been strongly discouraged.

Politics and policies are also significant influences in determining who can enter the STEM space. Legislative decisions, educational funding, and the policy frameworks will affect your access to quality STEM education and resources. Also, we cannot ignore socioeconomic factors; financial constraints are one of the main limiters of opportunity for marginalized groups, which is reflected not just in access to advanced learning tools and extracurricular activities but even basic STEM education.

The book advances from diagnosis to prognosis, mapping trends and suggesting ways through. Studying our successes through past models and strategies to effectively diversify, we can deduce best practices that can be copied and expanded. Inclusive curricula, mentorship programs, and outreach initiatives have all been found to be valuable in this regard. There needs to be an ecosystem in place between educational institutions and the industry that ensures diverse talent has a clear path from early education through professional development.

An important element of this narrative is the rapid pace of technological change. Science has a reputation for moving so quickly that marginalized communities can only fall further behind. But how do we make sure these groups are not always behind? The missing piece, many feel, is a matter of proactive inclusion. To incorporate diversity into the framework of STEM initiatives, not as appendages. With an emphasis on inclusivity, it becomes possible to seed sustainable and meaningful change.

The foundation of diversity in STEM begins with education. We need to reimagine (and redesign) our educational pipeline, and we must do it from K-12 up through higher-ed. These changes might include curriculum reforms that feature various scientists and engineers, interactive teaching practices that align with diverse learning styles, and mechanisms to help students from underrepresented groups. Exposure to STEM careers early on in life, along with similar role models who have pursued and succeeded in these fields, could encourage students to see themselves as future researchers or innovators.

The industry also has a major role to play in this. Businesses need to do work to diversify their pipelines and teams, hiring and retention strategies, workplace culture inclusivity efforts, and ongoing professional development opportunities. They must begin building a more inclusive environment by implementing employee resource groups, diversity training, and equity-based pay structures.

Again, it is the socioeconomic dimension of this argument that complicates things even further. Working to overcome the financial obstacles that impede entry and success in STEM fields is essential. Affordable educational programs, scholarships, and grants can help level the playing field. Public–private partnerships (PPPs) can build pathways into underserved communities as long as financial burden does not press out the creative and innovative.

Also, we highlight some successful international models and initiatives for promoting diversity in STEM that can be made use of in the United States. Using comparative analysis yields useful information and shows that the problems are indeed global, but so are the solutions. Through embracing the diversity of experiences and approaches, we can mold a more inclusive and exciting STEM world.

Overall, this book seeks to shed light on the diversity issues faced by STEM as well as to provide substantial solutions. We hope to address the complex challenges and avenues of approach in our strategy to promote an environment for everyone to succeed in STEM, wherever they come from. We hope that this book sparks conversation and serves as a model to show how we can work together for the day when diversity in STEM is no longer an ideal but a fact.

1

Historical Psychological Oppression of Marginalized Youth

1.1 Obstacles for STEM Learning for Marginalized Youth

Barriers to Science, Technology, Engineering, and Mathematics (STEM) education for marginalized youth have long existed and continue to shape the educational landscape. Understanding these barriers requires a deep dive into the historical and psychological factors that have contributed to the exclusion of certain groups from quality STEM education. These barriers are not relics of history but modern-day obstructions keeping some of our most underserved youth from opportunity. By making access to STEM resources inequitable, then only making STEM education possible by overcoming the exclusion practices within educational institutions, inequality is compounded. By identifying and overcoming these barriers, an educational environment that is inclusive and equitable for all students can be formed.

In this chapter, we will delve into different aspects of the challenges that underprivileged youth encounter in seeking STEM education. The first section provides a history of exclusion and details how long-standing policies, such as segregation, have created differences in educational resources and opportunities. The chapter then explores institutional practices like tracking and biased counseling that produce more obstacles for underrepresented students. In this section, we will investigate how access to STEM resources (e.g., technology and extracurricular programs) may serve as important determinants of participation in STEM fields. Lastly, the long-term consequences of such barriers will be presented, and the chapter will show that early exclusion of students from STEM education based on their socioeconomic status leads to less diverse persons in professions working within these fields.

1.2 Marginalization of STEM Education Groups

There have been systemic barriers that have historically restricted marginalized youth from STEM fields creating the existence of inequity. One of the most impactful factors is the historical barriers that have led to the exclusion of these groups

DOI: 10.1201/9781003453581-1

from quality STEM education. Over the decades, various policies and key events have contributed to this disparity. For instance, during segregation in the United States, African American students were often relegated to poorly funded schools with limited resources. These institutions lacked advanced STEM courses and qualified teachers, which created a significant educational gap. The ramifications of such exclusions are not just confined to the past but continue to influence the present opportunities available.

School administrative policies such as the sending and the sorting of students into different levels of educational paths based on perceptions of ability have placed marginalized youths at a particularly high disadvantage. Students from underrepresented backgrounds tend to be sent into less challenging STEM courses, reducing the likelihood of exposure to rigorous STEM curricula. For example, counselors may harbor some subconscious biases and discourage minority students from pursuing challenging STEM subjects and steer them toward less demanding tracks. This not only limits their educational experiences but also affects their long-term academic and career choices.

The other major obstacle faced by the marginalized youth was that of the access to resources. In quite a few of our underserved communities, the schools are devoid of any materials or equipment necessary for an effective STEM education. Underfunded and obsolete labs may deprive students the hands-on experience they need to guide their interest in STEM fields. This is only made worse by technology gaps, where students without the newest of computers and fast internet speeds struggle to compete with classmates in wealthier areas. Low-income schools in turn typically lack extracurricular programs, such as science clubs, robotics teams, and coding camps to generate or nurture a strong interest in STEM among their students. Solving that will require a range of approaches, from more money to community partnerships and focused efforts to ensure all students have equal access to STEM projects.

Recommendation: To establish such a relationship, schools and communities need to work together to close the resource gaps; additional resources could be donated by local businesses and universities, such as equipment and volunteer time for extracurriculars, and internships with real-world STEM experience.

Additionally, the downstream effects of premature exit from STEM education can endure for disadvantaged youth, shaping their ambitions and career paths over time. But when certain groups of students are continually denied good quality access to education in these areas, it is off-putting and does not encourage their interest and belief in their capabilities. This lack of impact accumulates, so that even if these students make it to institutions of higher education, they may experience impostor syndrome at times where their White peers have a heightened confidence and persistence in STEM fields, attributing to a higher sense of inferiority or dropping out. As a result, the current STEM workforce is overrepresented by a single demographic group, which inhibits innovative thinking and ability to cater to all within inclusive environments.

1.3 Perpetuation of Stereotypes About Marginalized Groups' Abilities in STEM

Stereotypes regarding the capabilities of marginalized groups have a profound impact on their engagement in STEM education. This section examines how these stereotypes manifest and influence various aspects of the educational experience for underrepresented students.

"Stereotype threat" is a concept that illustrates the destabilizing effects that negative stereotypes have on student performance and engagement. It occurs when individuals fear they might represent negative stereotypes about their social group, which can lead to anxiety and reduced performance. In the context of STEM education, students from marginalized backgrounds often face stereotypes suggesting they are less capable in scientific and technical fields. This fear of fulfilling these stereotypes can cause significant stress and affect their cognitive resources, leading to lower performance and disengagement from STEM activities (Totonchi et al., 2021).

In fact, research suggests that stereotype threat harms students most when they are at key points in their educational progress. For example, one longitudinal study found that perceptions of stereotype threat among underrepresented minority students grew over four years in college and were related to decreased science self-efficacy, intrinsic value, and attainment value (Totonchi et al., 2021). Taken together, these findings highlight the insidious nature of stereotype threat in undermining motivation and persistence in STEM fields.

This means, beyond stereotype threat itself, STEM curricular content and structure can reinforce stereotypes about the capabilities of minority groups as well. For some students, STEM curriculums lack relatable representation and ignore the inclusion of diverse scientists. So, without access to diverse role models in the first place, students begin to falsely imagine that one cannot succeed at STEM if they are not from a specific group or other. This lack of representation can limit their sense of belonging and potential within STEM. It may beg the question, "if the only successful people they see in these fields do not look like them, why should they believe that they could be successful too?"

Standard pedagogical approaches in many STEM subjects may inadvertently favor students who fit the stereotypical image of a successful scientist or engineer. Traditional pedagogical practices, focused on rote learning and competitive settings, might disadvantage students from underprivileged backgrounds who would instead benefit more if offered greater collaborative and context-based learning opportunities. Nevertheless, the iterative feedback loop of underrepresentation and stereotype reinforcement will persist in the absence of curricular reforms that correct for said biases where they manifest.

Peer perceptions also play a significant role in shaping students' engagement with STEM education. The attitudes and expectations of peers can have a strong impact on how students see themselves academically. Group dynamics in education often reinforce the wrong stereotypes. For instance, during group work or class interactions, minority students may experience their contributions to be marginalized and their participation devalued, which can also diminish their resolve to continue in STEM pathways.

Also, such comparisons may further marginalize students via peer pressure. In the presence of would-be peers who they think are better at or more inclined toward STEM, these students may come to believe that they just aren't cut out for it. This is especially true in many competitive academic institutions where "winning" has a very narrow and predefined meaning as measured by standardized achievement. Consequently, peer settings become an incubator of stereotypes instead of being a place that encourages the growth and safety of everyone.

Media representation significantly influences career aspirations among marginalized youth in STEM. Media representations of STEM tend to be one-sided, giving only a limited and unvaried perspective on who can pursue a STEM career. Pictures of scientists, engineers, and mathematicians in films, television programs, and advertisements are largely of White men. The problem is that if these are the only images of STEM we are shown, and for many people they are, then we create this myth that STEM careers belong to a special type of person.

If marginalized youth do not see themselves reflected in STEM roles, they are less likely to consider those as paths for their own future. In contrast, more positive and varied depictions can evoke aspirations and serve as examples for students to see themselves in similar roles. Media can break stereotypes and open the eyes of young people to what they think is possible. Showing a wider range of role models and success stories across the media, reflecting on all corners of society, can be key to breaking down stereotypes and opening up STEM fields.

To counteract the negative impacts of media representation on career aspirations, it is important to provide guidelines for creating more inclusive and representative media content. First, media producers should prioritize diversity in casting and storytelling, ensuring that people from marginalized backgrounds are depicted in positive and complex STEM roles. Second, collaborations with organizations dedicated to promoting STEM education among underrepresented groups can help create authentic and inspiring narratives that reflect the real-world diversity of STEM professionals. Third, media literacy programs can equip students with critical thinking skills to analyze and challenge stereotypical portrayals, empowering them to seek out and demand more inclusive content. In order to break these barriers in STEM education, the media industry can also play a major role by adhering to the guidelines that respect diversity and inspiring the global youth.

1.4 Inherited Societal Perceptions on Race and Gender in STEM Access

Many of these beliefs and expectations about race and gender in STEM are situated in our cultural assumptions. This puts half of the U.S. population in an uphill battle to combat societal beliefs and norms that suggest certain racial groups are inherently less interested or capable in the fields of STEM. This leads to the type of cultural assumptions that can make marginalized students internalize limiting beliefs, which in turn decreases their confidence and interest in pursuing STEM education.

For example, the stereotype that Asian students are naturally good at math and science may place an enormous amount of pressure on these individuals or may reinforce negative stereotypes about other racial groups. Marginalized students, particularly those from Black and Hispanic communities, may feel that they do not belong in STEM fields due to widespread societal prejudices that downplay their intellectual abilities. This not only affects their self-esteem but also impacts the opportunities they are offered, possibly limiting their exposure to advanced STEM coursework and extracurricular activities.

In addition to cultural assumptions, social stereotypes about intelligence further create barriers. Traditional definitions of intelligence in STEM fields often prioritize analytical and logical reasoning over other forms of cognitive abilities. This narrow perspective excludes other expressions of intelligence that might be more prevalent in different cultural backgrounds. For example, community-based problem-solving skills and practical knowledge, which are highly valued in some cultures, are often overlooked in mainstream STEM education.

This is a key step in breaking down archaic belief systems and building an intersectional space within the STEM world. Identifying the myriad ways to be intelligent will allow an opportunity for many students to access, as opposed to shut them out of, a successful academic pursuit. This greater definition of intelligence should be the model for teaching strategies, evaluating standards, and classroom fluency. Integrating project-based learning and group problem-solving can validate and leverage all individual intellectual talents.

Gender-specific barriers present unique challenges for marginalized girls in accessing and succeeding in STEM education and careers. Cultural stigmas may discourage girls of color from pursuing interests in STEM by promoting the idea that these fields are "not for them." This may start as early as elementary school and continue through higher education and also into their professional careers. Studies have shown that even when girls show an initial interest and aptitude in STEM subjects, they often receive less encouragement from their parents, teachers, and peers compared to their male counterparts.

These gender-specific barriers are not just limited to external perceptions. Internalized gender roles can lead girls to doubt their abilities and question their interest in STEM. Addressing these challenges requires a multifaceted approach that should include mentorship programs, representation of female role models in STEM, and their history being talked about, highlighting contributions of women in these fields. Schools and institutions need to create environments where the girls feel supported and encouraged to pursue STEM without facing gender-based discrimination.

Guidelines for supporting marginalized girls in STEM should include creating safe spaces where they can express themselves without judgment, providing access to female mentors in STEM professions, and ensuring that educational materials and activities are inclusive and representative of diverse contributions to the field. Design programs aimed at building confidence and resilience, such as girls' coding clubs and science camps, can also play a significant role in breaking down gender-specific barriers.

Additionally, by framing the issue of diversity in STEM through intersectionality, we are better able to understand the multiple levels of complexity at play that serve as barriers for marginalized youth of color. The intersections of race, gender, and social class add further layers of problems, as they create a multiple disadvantage situation for the people who turn out to be at their convergence. A low-income Black girl finds herself fighting discrimination based on race *and* gender along with economic disadvantages that keep her from a decent education and resources.

Understanding intersectionality is essential for addressing these multiple disadvantages. Policies and programs aimed at increasing STEM participation must consider the intersecting characteristics of students and tailor solutions accordingly. Providing scholarships and financial aid, offering flexible learning schedules for students who may need to work part-time, and creating awareness about the unique experiences of these students are vital steps toward creating an inclusive STEM environment.

By taking an intersectional approach, educational institutions can work with community organizations that have experience in catering to the needs of marginalized groups. A collaborative network between schools, local businesses, and community leaders must aim for a holistic approach that addresses the wide spread of barriers students face. Moreover, integrating intersectionality into educator preparation programs can also provide educators the tools to identify and combat more covert biases in the classroom.

1.4.1 Youth and Minorities in STEM: The 21st-Century Landscape

We all know how the face of the world has been changing with the introduction of technology. But young, marginalized minorities seeking to break into careers in these fields face long-standing barriers stemming from historical, cultural, and systemic inequity. Breaking down these barriers requires not just awareness but concerted efforts to expand access and encourage participation.

1.4.1.1 So, What Is the State of STEM?

Underrepresentation of minorities in STEM education and professional industries persists. Despite greater awareness of these disparities, the representation of Black and Hispanic people in the STEM workforce does not match their overall share of the labor force (Fry et al., 2021). As discussed in a number of studies, members of these groups are significantly less likely to attain degrees in STEM fields than are those from other parts of society, which is leading to sustained underrepresentation in career paths such as computing and engineering.

The data also show that, as of 2018, Native Americans, Native Hawaiians, Pacific Islanders, and those from multiple racial categories earned just 4% of bachelor's degrees and 3% of advanced degrees in science and engineering disciplines. This low degree attainment is symptomatic of larger systemic inequities in the STEM pipeline available to these communities. Women have surpassed men in obtaining undergraduate and advanced degrees overall, but women are underrepresented in engineering and computer science (Okrent & Burke, 2021).

This is aggravated by existing disparities and inequities in access to STEM resources. Structural inequities of place and socioeconomic status keep many minority students from accessing high-quality STEM education and resources at an early age. This educational deficiency impacts minority students accessing advanced STEM disciplines, resulting in limited career options to achieve upward movement within the fields of STEM.

There are also major areas of concern when it comes to representation and role models for STEM industries. Role models play a key role in stimulating youth to aspire and follow the pursuit of STEM-based careers. Unfortunately, the lack of diverse representation in the highest echelons of these fields can deter future applicants who do not see someone like themselves. Moreover, initiatives to raise awareness and spur interaction are necessary to generate interest and maintain commitment from students from underrepresented backgrounds.

Whether explicit or implicit, institutional biases persist in academic and corporate contexts. Such biases are manifest in hiring processes, retention policies, and promotion practices that systematically disadvantage minorities. But even when minorities gain access to STEM jobs, they have been shown in studies to contend with significant barriers to mentorship and sponsorship opportunities, which are critical for upward mobility and leadership development. Institutional cultures can also unwittingly reinforce feelings that make minorities feel that their contributions are neither appreciated nor valued.

Making STEM programs more inclusive is a critical component in combatting these issues. Institutions need to provide a range of support services and consideration for the policy needs of marginalized communities. This ranges from scholarship efforts or mentorship programs to initiatives in the form of community outreach where students and professionals working in related areas can connect with each other. Concurrently, a congruence of the curriculum with culturally relevant pedagogies can result in a more conducive learning and developmental environment for students.

1.4.1.2 Obstacles to Access and Representation

In the 21st-century STEM fields, there are opportunities and challenges for underprivileged youth and minority communities. These obstacles are pervasive and entrenched, tied to cultural, economic, and structural realities that need careful study to tackle. These groups have notable cultural and socioeconomic barriers to entry into STEM fields. Many of them are raised in circumstances where access to high-quality education is more limited, which can lead to disadvantage gaps in the foundational knowledge and skills essential for reaching STEM fulfillment. Economic restrictions can pose additional barriers to participation, such as when families may not have access to pay for educational things for science and technology. In addition, cultural ideas about careers may deter young people from pursuing STEM pathways, particularly if a career within STEM is viewed as out of reach or unattainable in a community context (Jantzer et al., 2021).

Another significant barrier lies within educational systems with implicit biases and stereotypes. These biases can lower expectations or result in differential treatment by educators, contributing to a self-fulfilling prophecy in which students internalize negative stereotypes and experience lower levels of engagement (Thiem & Dasgupta, 2022). Such an atmosphere erodes a student's confidence, sense of belonging, and desire to stick with challenging courses like math and science.

Educational inequity is further exacerbated by the uneven distribution of funding across regions; the disparities in the quality of education available are often staggering. Many public schools that primarily serve minority populations are critically underfunded, meaning students have far less access to advanced placement courses, new materials, and adequate facilities for STEM education than their counterparts. The absence of investment in out-of-school time sends the signal that the education of those overlooked students is not valued and severely restricts their exposure to the critical skills necessary for careers in STEM fields. These inequities persist after primary and secondary education, expanding into areas of higher education, where limited funding inhibits programs created to support underrepresented minority (URM) students in achieving STEM degrees (Jantzer et al., 2021).

For a first-generation college student, it's a journey with a few extra hurdles to navigate on the path to a STEM career. Because these students might not have family members who experienced higher education systems to help guide them, they are left to become familiar with application processes and financial aid and the intensity of academic life on their own. In many cases, with little or no connection to networks of older mentors or role models in fields of similar interest, these lower-income students feel isolated and overwhelmed. And by the time they enroll, they may encounter dominant academic cultures that do not recognize and/or accommodate the various backgrounds and experiences that they bring, reinforcing alienation and discouragement. In STEM, they experience attrition rates higher than others, their counterparts (Thiem & Dasgupta, 2022), that do not allow them to flourish and keep pursuing their studies in STEM fields.

Addressing these challenges will substantially involve intervention strategies that reflect the unique contexts of marginalized youth and minorities. For minority students, mentorships can serve as a great benefit, which supports them through their journey through both the academic and the working world. Mentorship programs match students with professionals in their field who can provide motivation and real-world advice on how to overcome hurdles. These programs have stated that retention and changes to that have been due to a sense of community and belonging.

1.4.1.2.1 Practices That Advance Inclusion and Diversity

In an effort at increasing diversity within STEM, many programs have been set in place to ensure underrepresented youth and minorities are properly represented. These efforts focus on removing barriers and building communities where underrepresented groups in STEM can flourish.

Policies from their governments are the tool that sets the tone for greater representation in STEM. There has been a flurry of proposed legislation, both federal and local, to further this end. Typically, these policies emphasize increasing funding for

STEM programs in underrepresented communities, promoting scholarship opportunities for minority students, and fostering partnerships between academic institutions and industry stakeholders. This collaborative approach is critical for making STEM education and careers accessible for everyone, regardless of their background. Moreover, such policies not only aim to bridge school-to-work transition gaps but also to lessen systemic inequality by establishing inclusive norms and creating space for multiple perspectives.

One of the major ways nonprofit organizations are improving diversity in STEM is through mentorship, educating and even sponsoring young learners. These organizations fill the gap between schools and students with resources, guidance, and networking opportunities. Organizations such as Black Girls CODE and the National Society of Black Engineers (NSBE) create environments in which future engineers or scientists connect with professionals who offer important career guidance and support. Such mentorship programs allow students to learn more about potential career paths, and provide a source of encouragement as they face challenges in predominantly homogeneous fields.

The science literacy of underrepresented populations is also boosted through community-driven efforts. Grassroots efforts are more customized to the local population, where they can base their focus on early childhood education and community involvement. STEM workshops and other activities designed to inspire people in science from the earliest point in their lives are delivered through hubs within libraries, community centers, and local schools. By engaging peers, parents, educators, and local leaders, these programs build holistic support systems that feed students' curiosity and drive. Furthermore, incorporating STEM into community programs cultivates learning opportunities outside of the classroom that can encourage students informally but helpfully.

Corporate responsibility initiatives create additional support by promoting inclusive hiring practices and sharing the value of diversity in the workplace. Most companies understand that diverse teams help them to innovate and be more competitive. As a result, they initiate programs that focus on hiring people from diverse backgrounds, especially those that have been underprivileged. The internships, apprenticeships, and training programs created by these employers can help personalize STEM skills development for potential employees to succeed in STEM positions. Such corporate outreach programs are about more than providing opportunities for college students; they also represent cultural change in the workplace, where the thought goes that organizations need to create exposure and celebrate contributions from everyone, regardless of their position on the organization chart.

STEM community programs need to be designed and implemented with careful consideration so that not only educators become empowered by effective content delivery models but also community leaders are empowered to integrate STEM into their curriculum. Providing professional development workshops for teachers and trainers empowers educators to teach STEM subjects in novel ways that are engaging for young learners. The role for public and private sector collaboration can magnify these efforts, resulting in environments with abundant resources for technology and innovation.

Government initiatives and grants support endeavors by providing financial resources needed to sustain and expand STEM inclusion programs. Grants such as those offered by the National Science Foundation (NSF) foster research and program development aimed at increasing participation among underrepresented groups. These funds enable organizations to implement evidence-based practices and give the ability to measure their impact, ensuring that successful models can be replicated and scaled up nationally.

1.4.1.2.2 Educational Innovations and Their Impact

The influence of educational reforms on STEM access for marginalized youth and minorities in the 21st century is multifaceted. One of the most critical areas impacted is the implementation of curriculum adaptations tailored for diverse learning needs. In recent years, educational institutions have recognized that a one-size-fits-all approach to learning does not suffice, especially in subjects as complex as those in STEM. Schools are now incorporating different instruction techniques, which allow educators to cater to individual learning styles and abilities. These adaptations can take various forms, including project-based learning, which enables students to engage with real-world problems and develop critical thinking skills, and flipped classrooms, where students are introduced to new content at home and practice working through it during class.

Supporting these curriculum changes is the utilization of technology, a vital tool in bridging educational gaps faced by students from marginalized communities. Digital platforms offer interactive, entertaining avenues to augment STEM learning. They provide simulations, virtual labs, and other resources that may be logistically or financially inaccessible otherwise. Technologies such as online science laboratories enable students to do at-home experimentation in a virtually real environment, which is particularly helpful in schools that do not have the required physical setups. Also, artificial intelligence (AI) and other technologies can customize personalized learning experiences, differentiating challenges and supporting factors in real time based on assessments of student performance.

There are institutions that have revamped their extracurricular activities to engage a wider audience in STEM fields. Through clubs, competitions, and mentorship opportunities, these schools create pathways for students to explore STEM within supportive environments. Also noticeable is that school–organization partnerships for apprenticeships, internships, and so on allow students firsthand experience and exposure in STEM career fields.

STEM education is also being democratized through online learning platforms. Websites such as Khan Academy and Coursera provide free or affordable courses in every aspect of STEM education, opening up the door to high-quality education for students around the world, no matter their income or geographic location. As the world transitioned to online learning during COVID-19, a technology-adaptation of the education system became one of the most significant factors enabling educational resilience. Although this rapid transition came with struggles, it also underscored the potential for digital platforms to bring services to underserved populations. Learning remotely is especially important for students with transport difficulties or those living in remote areas far from educational facilities.

Furthermore, these platforms are often inherently equipped with a variety of features that cater to different learners—for instance, subtitles, transcripts, and variable speed of playback, permitting students to learn at their own pace. Recent research indicated that the COVID-19 lockdown also affected students' self-efficacy and motivation in STEM subjects as a result of the loss of face-to-face interactions with peers and teachers (The Impact of Online STEM Teaching and Learning during COVID-19 on Underrepresented Students' Self-Efficacy and Motivation I NSTA, n.d.). That said, the ability to study online—and with greater flexibility—has much promise in easing traditional barriers to entry into STEM areas.

Collaborations between the education sector and tech companies are increasingly important to create resources and tools that can augment STEM learning. Companies also often bring the latest technology, financial resources, and industry knowledge that can enhance education experiences. Effective partnerships can result in programs where tech companies adopt schools, providing mentoring, workshops, and donations of equipment. Such collaborations foster an environment where students can imagine themselves in future STEM roles, aligning with broader institutional goals to increase diversity and inclusion (D&I) within these fields.

As educational reforms continue to evolve, there is a need to pay attention to the needs of marginalized groups so that access to quality STEM education is equitable. Maintaining such an allied environment would require significant effort from both educators and policymakers to more systematically combine their efforts, address the socioeconomic and cultural gaps, and create an ecosystem where every student can thrive. By applying targeted changes to the curriculum, thoughtful applications of technology, and innovative PPPs, we can take the first steps toward a more inclusive STEM workforce and enrich the field with a tremendous number of different perspectives and ideas.

1.4.1.2.3 The Future Pathways for Marginalized Youth in STEM

Emerging STEM fields are increasingly offering new career paths for minorities, reflecting the dynamic nature of STEM. Fields such as renewable energy, biotechnology, AI, and cybersecurity are experiencing rapid growth and present significant opportunities for diverse populations. These areas are not only pivotal in shaping our future but also hold immense potential for increasing representation among minority groups, which could lead to better economic outcomes and greater diversity in problem-solving approaches.

This makes it more important than ever to lay the groundwork during childhood education to encourage an interest in STEM and prepare the young for exciting career opportunities in the field. Studies have indicated that children who are introduced to innovation and math ideas early on are able to develop superior tractability and a deeper love for STEM subjects. Introducing STEM through the school curriculum in the earliest grades can inspire curiosity and confidence in students, allowing them to imagine themselves as future scientists or engineers. For example, encouraging more STEM-related events in the community and outreach to underserved families can bring exciting hands-on experiences to children who may not have access. However, collaborative initiatives that involve both schools and families have been shown to be a successful way to keep children engaged with their STEM education extended beyond the classroom and lays a foundation for long-term learning (Taylor, 2023).

University–industry partnerships can play a role in building pathways for marginalized students. These collaborations allow for resources, mentorship, internships, and practical work experience that connect the academic setting to real-world application. Through utilization of these relationships, institutions can provide students with a better perspective on realistic career expectations and industry demands. Universities such as UTeach at the University of Texas at Austin exemplify this collaborative model with programs that get future educators ready to inspire the next generation of STEM leaders (Palid et al., 2023). Through partnerships like these, students are exposed to the possibilities that exist across different STEM industries, and can then become equipped to enter into and succeed in the workforce.

Strategies for achieving equitable representation in STEM worldwide are crucial for nurturing a truly diverse and inclusive environment. To reach this goal, educational institutions must prioritize diversity, equity, and inclusion (DEI) initiatives. Focusing on breaking down barriers that have historically limited access for underrepresented groups. Equity-centered programs, such as STEM intervention programs tailored for minority populations, have shown promise in improving retention and graduation rates among minority students pursuing STEM degrees (Palid et al., 2023). Also, promoting inclusivity involves implementing policies that address implicit biases and stereotypes, ensuring that all students feel welcomed and valued within educational settings.

Government policies aimed at encouraging diversity in STEM fields can also play a vital role in transforming the landscape. Governments can start offering incentives for companies that support inclusive hiring practices or contribute to building local talent pools via scholarships and training programs. That can drive changes toward fair representation across industries. By aligning policy goals with industry needs, societies can create sustainable models that address underrepresented voices within STEM, allowing everyone to benefit from advancements in knowledge and technology.

To also achieve equitable representation in STEM, it is important to highlight the role of mentorship and community-driven efforts. Networks that connect minority students with role models and professionals who share similar backgrounds can significantly impact student's career trajectories. Through formal and informal mentoring relationships, students gain access to guidance, advocacy, and resources that help them overcome challenges unique to their experiences in STEM environments.

1.5 Wrapping Up

In this chapter, we have taken a deep dive into the many obstacles facing marginalized students in their efforts to prepare for STEM education. We have delved into the ways in which STEM divides are exacerbated by exclusionary practices and social stigmas. Both historical configurations such as segregation and institutional practices like tracking and biased counseling emerged as central features in the maintenance of inequities. The challenges have been further compounded by resource limitations in underserved communities. The good news is, recognizing these are the biggest challenges.

Furthermore, the old-boys network creates a ripple effect of generational disadvantages: students facing systemic barriers in their home regions will never know to aim big for their careers. Early lack of exposure to quality STEM education not only dilutes interest but whittles away at self-assuredness, making success in higher education less likely and reducing the number of STEM degrees. Creating that environment for everyone requires redefining intelligence, eradicating stereotypes, and screening resources to be more equal. Together, schools, neighborhoods, and media have an enormous role in ensuring representation and advancement. This is the winning strategy to make a path for an inclusive STEM environment.

2

Limited Pipeline Resulting in a Homogeneous STEM Environment

2.1 Looking at the Impacts of a Limited STEM Pipeline

Analyzing the effects of a limited STEM pipeline uncovers the numerous challenges marginalized communities face in accessing and succeeding in STEM fields. They contribute to significant underrepresentation, which has broad implications for both the individuals affected and society at large. Understanding these challenges is crucial for implementing effective solutions that promote D&I within STEM disciplines.

In this chapter, we explore the psychological barriers that obstruct engagement from marginalized youth, namely, effects of low expectations and a lack of varied role models. We also look at how bad classroom atmospheres and the reassurance of failure hurt student engagement. In addition, the chapter examines specific interventions and programs that aim to reduce these barriers through enhanced role models and faculty visibility, more inclusive classroom culture for students from underrepresented groups, as well as curricular strategies focused on stereotype threat reduction within culturally affirming teaching environments. We hope this full survey will further underscore the need for equity in STEM, capturing where our challenges lie and clearly showing us what we must fix to achieve that more equitable and complete pipeline.

2.2 Psychological Oppression of Grade School Youth Due to the STEM Divide

Psychological barriers are critical in determining the engagement of marginalized youth with STEM education. These barriers can take many shapes and sizes but together contribute to the underrepresentation and disengagement of these students in STEM fields.

One of the most significant psychological barriers to STEM education for marginalized youth is the effect that low expectations have upon them. Low expectations among educators, parents, or society as a whole can lead to feelings of incompetence in minorities. This sense of lower ability can result in reduced motivation and interest

DOI: 10.1201/9781003453581-2   14 to study STEM fields. Research suggests that the experience and achievement levels of various groups of students differ markedly not only based on their learning environment but also due to what is expected from them (Easterbrook & Hadden, 2020). When marginalized students take on these low expectations and feel unwelcome in high-achievement fields like STEM, the result could be to lower their participation or cause them to not persist in these arenas.

Another critical barrier is the lack of diverse role models in STEM fields. Role models play an essential part in inspiring students and providing them with examples of success that they can aspire to. Yet, the underrepresentation of women and some ethnic minority groups in many STEM fields contributes to lack of available role models. While a picture may say a 1,000 words, an absence of representation tells a lot more. For marginalized youth, it can lead to them interpreting this lack of representation as no one who looks like them or shares their backgrounds is able to be successful in STEM subjects and it causes isolation within the wider scientific community. This lack of representation devalues the importance students place on STEM careers, leading them to question whether these fields are welcoming or attainable for them.

Fear of failure in a predominantly homogeneous environment is another substantial psychological barrier. In STEM classes where the majority of students belong to a dominant group, marginalized students might feel heightened anxiety about confirming negative stereotypes associated with their identities. This stereotype threat can cause more stress and worse performance in the affected populations; thus, these students are less likely to take STEM courses at all. Stereotype threat has been clearly demonstrated to lead to a decrease in motivation and performance, especially among those who are highly invested in academic domains (Totonchi et al., 2021). The fear of living up to those stereotypes can be so palpable that some students do not identify with the academic domain altogether, meaning they no longer see success in that space as a relevant goal for their lives.

These experiences are compounded by negative classroom dynamics for marginalized students in STEM education. Classrooms where it is a priority to compete with one another rather than work together can isolate students who are already underrepresented in the classroom. Marginalized students often find that cutthroat competition flourishes in their fields of study, and such environments push them away from engaging with the material. These negative interactions with peers or instructors will serve to reinforce the feelings of inadequacy and isolation that already drive underrepresented minorities away from these fields.

The psychological barriers can be addressed by provision of targeted interventions in an educational environment that supports inclusion, encouragement, and positivity. Maintaining high expectations for every student, regardless of their background origin, can reduce the way in which our own mind politics may harm their potential. Visiting speakers, shining the light on role models from diverse communities, and connecting students to mentors who will nurture and guide them are all ways in which institutions can help to improve representation.

Additionally, fostering an inclusive classroom culture that values collaboration over competition can create a more welcoming environment for marginalized students.

Educators should emphasize teamwork and collective problem-solving, encouraging students to support each other in their learning journeys. By creating a classroom atmosphere where every student feels valued and included, educators can help mitigate the negative impacts of psychological barriers and promote greater engagement in STEM subjects.

Additionally, combating stereotype threat demands individual and collective action from educators and organizations alike. By becoming aware of and managing stereotype threat training for teachers can have the same beneficial effect. Developing curricula and teaching strategies that are culturally responsive and inclusive can also help reduce the anxiety and stress associated with being an underrepresented minority in STEM fields. In addition, hands-on projects, internships, and infusing STEM with real-world applications also afford students opportunities to gain confidence in that experience, which can increase student engagement and persistence within these pathways.

2.3 Why We Need More Efforts to Close Equity Gaps in STEM

Addressing equity in STEM education and the workforce requires targeted initiatives that consider the needs of marginalized communities. Despite numerous efforts, many current initiatives fail to reach their full potential due to gaps in outreach and overall effectiveness (Palid et al., 2023). For instance, while programs might focus on increasing statistical representation, they often overlook vital elements such as inclusion, sense of belonging, and persistence among students from underrepresented backgrounds.

Community-based approaches, which are tailored to local challenges, offer a more effective means of creating trust and interest in STEM fields. These initiatives, rooted in understanding the specific obstacles and opportunities within a community, not only increase participation but also build long-term commitment to STEM careers. By engaging local leaders, educators, and families, community-based programs can create an environment where students feel supported and motivated.

One key factor in maintaining student interest in STEM is longitudinal mentorship. Consistent mentorship helps students navigate academic and career pathways, providing guidance and encouragement over extended periods. There should be regular engagement through workshops, seminars, and industry collaboration that further supports this interest. For example, partnerships with local businesses and industries provide students with real-world experiences, making STEM subjects more tangible and relevant to their lives and future careers (Cobian et al., 2024).

But, the effective use of technology and access to resources have become critical elements in making STEM more accessible for all. With technology, students can work together on projects and use learning resources to fill in the gaps left by traditional classroom education. They have the option to learn from online platforms, virtual labs, and educational software linked to scientific equipment to make even complex concepts easy while breaking all barriers of geography or economy.

To be successful, these initiatives need to be designed with careful consideration of the unique needs and circumstances of the target communities. Outreach programs must be inclusive, ensuring that information and resources are accessible to all, particularly those who have been historically excluded from STEM fields. This means producing materials in multiple languages, using culturally relevant examples, and actively reaching out to underserved areas.

Second, these programs need to be evaluated continuously for impact. The feedback from participants, community members, and educators will inform the constant refinement of the initiatives to be responsive and impactful. Finally, beyond direct impacts on enrollment numbers, evaluations should also consider how these initiatives affect student retention, graduation, and success in STEM fields. These longer-term metrics will better help to identify which strategies are and are not having the greatest impact.

By providing tailored mentorship opportunities, involvement in community, and strategic use of technology, we will be able to create a more inclusive pathway toward STEM education. By taking on the enormous disparities in access and opportunity that already exist, we have a chance to enable an entire generation of underserved students to do incredibly well in STEM. Integrating these together into an effective strategy will be a game changer in working toward real equity in STEM education and the workforce.

In order to strengthen these points, consider a couple of examples. In a rural area, the program might collaborate with local schools and regional businesses to provide access to STEM projects, mentorships by working professionals, and internships. Such experiences can help to demystify STEM careers for students who may not have viewed them as within reach. On the urban initiative side, technology could connect young innovators with learners and resources on a global scale in a way that they would not normally access it within their locality.

2.4 Current Demographics of the STEM Pipeline Reviewed

To investigate the impact of a short STEM pipeline, we need to look more closely at the current state of play in these fields. This overview aims to address critical differences and emphasize areas to which we should pay attention, making a comprehensive assessment beyond the scope of an overview of this type.

When we look at demographic representation of who is in STEM, the statistics are pretty bad. As an example, women and ethnic minorities are underrepresented in a range of educational and professional levels. Women, who earn nearly half of all science and engineering degrees nationally, still only comprise approximately 28% of the STEM workforce (Ovink et al., 2024). But even though the Black and Hispanic share of the general population is ticking higher, Black and Hispanic professionals comprise only a fraction of the workforce in these sectors. If we identify these gaps, we can focus our efforts by closing the respective holes.

Analyzing the way trends have changed over time is essential to evaluating societal attitudes and educational policies regarding STEM participation. In the past, many groups, particularly women and minorities, had struggled with systemic barriers when it came to participating in STEM education and working in STEM careers. Historical data enable us to trace the change in participation rates over time due to particular policy interventions or shifts in societal attitudes. In the United States, perhaps most notably the introduction of Title IX, a federal civil rights law that prohibits sex discrimination in education and provides protection for athletes, has been linked to an increase in female college enrollment overall, including in STEM fields (Dost, 2024).

An analysis must fully acknowledge intersectionality and how people's multiple self-identities combine to structure the social positions they hold in STEM. That is to say, the barriers a Black woman in STEM faces are not just a function of the amount of racial and gender barriers experienced but uniquely defined as challenges borne out of the intersection of both identities. A combination of discrimination appears to influence the belonging and persistence within STEM fields for women of color (Ovink et al., 2024). Overall, having a better understanding allows us to build more effective support systems that accommodate and can actively address these needs within the STEM space.

In addition to analyzing current data, there is also a need for more robust data collection to better inform policy and action related to STEM equity. Current metrics often fail to capture the full scope of diversity within STEM, overlooking factors such as socioeconomic status, first-generation college status, and other relevant dimensions. Robust data collection methods are necessary to identify trends accurately and evaluate the effectiveness of interventions. Universities and research institutions should prioritize the development of analytics capabilities that provide granular insights into student demographics and outcomes. These tools can help tailor interventions to better serve marginalized communities and track progress over time.

Similarly, such patterns are observed when looking at how societal attitudes affect STEM participation more broadly. Who we think belongs in STEM fields is largely informed by cultural stereotypes and biases. Take, for example, the stereotype that scientists are mostly White men, which can dissuade underrepresented minorities from entering STEM fields. Initiatives aimed at combating these stereotypes, such as public awareness campaigns and representation efforts in media, can gradually change societal perceptions and encourage more diverse participation in STEM (Dost, 2024).

STEM demographics are also heavily influenced by educational policies. When the rubber meets the road, so to speak, policies that advance inclusive curricula and afford equitable resource access can close these gaps. Summer bridge programs and faculty mentoring have shown some promise in helping disadvantaged students succeed. Yet, these programs only work if they are developed from an understanding of the specific constraints faced by these students (Ovink et al., 2024).

Part of the emphasis on intersectionality is dedicated to emphasizing the need for conglomerate support structures. Programs such as mentorship pairing up minority students with mentors from similar backgrounds can have a huge impact on the improvements to feelings of belonging and retention for underrepresented marginalized groups in STEM. Moreover, the development of inclusive classrooms that embrace diversity while establishing strong support systems can help to combat stereotype threat and create a more hospitable environment for all students (Dost, 2024).

And last but not least, the imperative of national level information must not be understated. But without accurate and detailed data, policymakers and educators are working in the dark; they simply cannot see what needs to go where or how best to design interventions. Longitudinal studies that trace students from middle school into adulthood, for example, can give us a look at how different educational policies and programs play out over the long term. Disaggregated data, moreover disaggregating by specific subgroups within the marginalized communities (Ovink et al., in press), could expose more nuanced challenges and inform tailored solutions.

2.5 Challenges Faced by Marginalized Communities in Pursuing STEM Careers

There are many barriers that marginalized communities face when pursuing a STEM career. Among the largest is, of course, economic. Despite promising job prospects, the high price of STEM programs is a barrier for students from low-income backgrounds. Tuition fees, text books, lab equipment, and similar expenses might discourage some students from even commencing. Furthermore, a heavy debt load for students is used as a disincentive to keep people out of higher education in STEM fields. The financial constraints restrict the number of students who can study in these industries, furthering an underrepresentation cycle for such vital fields.

The lack of networks also contributes greatly to restrict opportunities for marginalized populations in STEM career paths. In terms of job placements, professional networks are powerful tools when it comes to job, internship, and career opportunities. But students from underrepresented backgrounds often have limited access to these networks Many professional associations and informal networks in predominantly White institutions do not adequately include or support marginalized groups. This exclusion hampers their ability to gain mentorship, insider knowledge, and job referrals that are critical to advancing in STEM careers.

To address networking disparities, it is essential to create inclusive professional organizations and networking events. Institutions should proactively reach out to underrepresented groups, offering scholarships to attend conferences and facilitating connections with mentors who can provide guidance and support. The development of relationships with STEM organizations that focus on the topics of D&I can also be useful in addressing such issues. Examples include the Society of Women Engineers (SWE) and the NSBE, both national professional organizations with specific programs designed to help marginalized communities in STEM.

For marginalized youth, the cultural barrier only makes finding and working toward STEM careers more difficult. In some communities, STEM pathways may be seen as not congruent with cultural values or even out of reach. These perceptions can be off-putting and deter young people from pursuing studies in STEM. However, this approach can be counteracted by linking STEM education more closely to community values and needs. This change can be accelerated by community outreach programs and culturally relevant teaching practices. Educators can engage community leaders and activate local role models from similar backgrounds to demonstrate STEM careers as being both meaningful and attainable.

Moreover, embedding STEM education into culturally significant projects can increase its appeal. Among them are attracting indigenous students to environmental science by incorporating traditional ecological knowledge into the curriculum. This could also generate interest and engagement by illustrating how engineering innovations can solve problems pertinent to the community. To creating a STEM environment that values and acknowledges diverse cultures, many schools and educational STEM programmers will need to do more, to include learning how culture is experienced by many people as well as with specific cultural groups.

Often times, the main barrier for fostering more inclusive and equitable STEM teams is not related to qualifications but instead to workplace challenges such as microaggressions and exclusionary practices. What many employees experience, in some variation, at work are microaggressions—those everyday verbal or nonverbal communications or small behaviors that denigrate others. These experiences are demoralizing and further alienate disadvantaged employees. In addition, exclusionary tactics like not being invited to important meetings or involved in key decision-making processes during career development build and compound low self-worth.

Organizations need to deliver impactful DEI training in order to generate solutions for workplace dilemmas. All employees should be trained in such programs to understand the significance of diversity and how microaggressions can harm others. Second, all companies should have a clear and firm policy to report and tackle discrimination. One of the worst things is to have a bunch of employees that feel like they do not have the autonomy or freedom to come forward about what they are experiencing and create an environment where they can thrive. Mentorship programs or affinity groups are also critically important for community, as they support those often left out of traditional work networks to find their place and advance in their careers.

Encouraging more opportunity for growth is another vital strategy in keeping diverse talent. Underrepresented employees, who receive unfair evaluations and lack access to leadership roles, often do not acquire promotions. It is important for organizations to make their promotion standards clear and fair. Standardizing performance reviews can reduce subjective biases and help in identification and development of a pipeline of high potential talents from underrepresented groups. Organizations that invest in leadership development programs specifically for marginalized employees can bridge the gap and cultivate a more inclusive leadership pipeline.

In summary, achieving equity for STEM careers is a complex problem where several factors have to be taken into account not only in education but also in our culture. Barriers to the access of wealth can be minimized through available scholarships, grants, and priced access to resources These systematic differences in networking call for intentional work to create space for all types of professionals in inclusive networks and mentorship pipelines. STEM education that is grounded in community values and culturally responsive teaching can mitigate cultural barriers. Workplace-specific challenges, like microaggressions and exclusionary practices, require robust DEI initiatives that work in concert with the support networks to provide an inclusive and equitable environment.

2.6 Successful Pipeline Intervention Programs

Highlighting best practices from programs that have worked to increase diversity in the STEM workforce is key to helping us address some of the structural barriers faced by marginalized communities and enable everyone no matter where they come from. The aforementioned case studies demonstrate how companies have gone about targeting and reeling in diverse talent, as well as the milestones these programs have achieved.

The Meyerhoff Scholars Program at the University of Maryland, Baltimore County, (UMBC) is one shining example. The program, initiated in 1988, has helped boost the number of minority STEM degree recipients. It is characterized by an aggressive recruiting campaign designed to speak directly to underrepresented minorities who are high-achievers academically. It features holistic support options such as mentorship, summer bridge programs, and financial programs to enhance academic and social success of first-generation students. An additional productive initiative is the National Institutes of Health's (NIH) Maximizing Access to Research Careers (MARC) U-STAR Program. The grant program was established to support the research training and career development of individuals from populations underrepresented in biomedical sciences who would be entering initial doctoral degree programs or postbaccalaureate research education experiences at institutions that have developed graduate partnerships. Participants have many opportunities in research, seminars, and professional development courses that make them very well-prepared for success in graduate-level education and research careers.

We can measure the real impact of these programs through a number of data points. UMBC's Meyerhoff Scholars Program, for example, has a stellar graduation rate of more than 90% in STEM fields among its participants. Further, many graduates have gone on to pursue graduate studies still widening the talent in academic and research career paths. MARC U-STAR, meanwhile, has successfully maintained higher retention and increased numbers of minority students matriculating toward graduate programs. Participants are more likely than nonparticipants on the same campus to graduate and pursue STEM careers, according to a study using data from multiple sources (Whittaker & Montgomery, 2012).

The following successful programs provide a few lessons learned from the highlighted examples, lessons that speak to an overall need for flexibility and collaborative, iterative change in the way we approach diversity in STEM. With regard to flexibility in program design, first, this allows for institutions to design programs that fit the needs of the students. Initiatives like the Meyerhoff Scholars and MARC U-STAR realize that they have to evolve with the times, receiving continuous feedback as to what is working and what is needed. Second, these programs are made more effective through collaboration of a variety of stakeholders (faculty, industry partners, community organizations). In many cases, students get more resources expertise and networking out of successful initiatives due to the partnership involved in an initiative. Third, an iterative approach is crucial. Regular monitoring and improvement of the program mean that new problems are identified and, more so, success remains relevant over time.

An example of adaptability is seen in the Louis Stokes Alliance for Minority Participation (LSAMP) program, which operates across multiple institutions to increase the number of STEM degrees awarded to underrepresented minorities. The program adapts its strategies based on the unique circumstances and challenges of each participating institution. Through regular evaluations and adjustments, LSAMP maintains its effectiveness in supporting minority students.

Another essential aspect of these models is an integration involving collaboration. Such collaborations between UMBC and local industries offer Meyerhoff Scholars internships, research placements, and experience with the applications of their in-class pursuits. This partnership will prepare the next generation of students for their future careers and create a sense of unity, belonging, and professional identity in STEM.

It is an iterative approach that embodies the continuous evaluation and improvement cycles within these programs. Adjustments are made to enhance program outcomes based on feedback from participants, faculty, and external reviewers. For example, the MARC U-STAR Program routinely changes its curriculum and cocurricular activities due to participant input and trending changes in industry. This iterative process ensures that the program continues to be applicable and viable in preparing minority students for STEM careers.

For scaling adopted models, future recommendations include policy shifts and increased funding for pipeline programs. There are some approaches that policymakers could consider in rewarding institutions that have a commitment to enhancing representation and inclusiveness of diversity within STEM. This again requires attention to where grants and funding are targeted, prioritizing approaches that have demonstrated efficacy to produce effects. Also, scaling successful programs to additional institutions, like those that serve many URM students, can help address disparities more broadly.

Scholars specifically urge setting out standards for both success and failure for policies intended to increase diversity. Evaluation criteria should be standardized so that institutions could know the best practices, bring them to light, and share able strategies within the broader higher education ecosystems. In addition, promoting a culture of learning and responsibility will ensure that diversity objectives are met in the long run.

It can also be of greater benefit to just implement mentorship programs on a larger scale. One of the success factors for diversity initiatives is mentorship. Growing formal mentorship programs that pair students with professionals in their field provides the necessary guidance, encouragement, and inspiration. These relationships help bridge gaps in professional knowledge and networks that often hinder underrepresented students' progress in STEM.

The next suggestion for the future is to advocate for inclusive teaching techniques. To improve student engagement and retention, one of the most effective steps we can take is training our faculty to embrace a culturally responsive pedagogy and establish inclusive classroom environments. Beyond technology, institutions must focus on faculty buy-in and invest in professional development that gives educators the skills and knowledge to effectively support diverse learners.

Also, what is of importance is access to research for underserved students!! A stipend, research assistantship, and travel funding for conferences can promote participation in STEM. Providing experience in research at an early age can inspire interest and increase the sense of belonging among aspiring scientific researchers.

2.7 Summary and Reflections

This chapter delved into the psychological and structural barriers faced by marginalized communities in STEM fields. We explored how low expectations, lack of diverse role models, fear of failure, and negative classroom dynamics contribute to these challenges. Creating solutions such as holding high expectations, fostering inclusivity, mitigating stereotype threats, and promoting collaborative learning environments were discussed to enhance engagement in STEM education. Additionally, the necessity of targeted community-based initiatives, personalized mentorship, and use of technology in bridging equity gaps was emphasized.

We also reviewed current demographic trends in STEM, highlighting underrepresentation among women and ethnic minorities, and the compounding effect of intersectionality. Some effective interventions were outlined, such as culturally relevant teaching and comprehensive data collection, to better understand and address disparities. Example programs were discussed that demonstrated successful efforts, like the Meyerhoff Scholars Program and LSAMP, showcasing adaptable strategies and collaboration efforts that support diverse talent. Future recommendations would include policy changes, expanded funding, inclusive teaching practices, and enhanced research opportunities to create a more equitable and inclusive STEM landscape.

3

The Impact of STEM Initiatives

3.1 Expanding the STEM Field Through Educational Outlets

A single broad intervention is unlikely to remedy the low diversity in the STEM field; specific educational initiatives must be directed at a variety of levels by starting early and continuing through higher education. By engaging students in hands-on activities outside traditional classroom settings, such as science fairs and coding clubs at the K-12 level, programs can effectively spark interest and dismantle existing biases about who can excel in STEM fields. Practical experiences build a strong foundation and inspire a long-term interest that follows students into higher education and beyond.

This chapter delves into the roles played by K-12 schools, colleges, and vocational training in creating diversity in STEM. It evaluates how early exposure builds foundational skills and explores how college programs can provide crucial resources like scholarships and mentorships to enhance workplace readiness. The chapter also highlights the level of undergraduate research and company partnerships available that set tuition-based colleges up to become feeder institutes for technical positions. Moreover, vocational training provides hands-on education that matches the requirements of the job market, thus leading to a job immediately and skills acquisition. Together, these multipronged initiatives, create a sustainable pathway to the empowerment of diverse Stem talents and inclusive growth.

3.2 K-12, College, and Vocational-Level STEM Initiatives

This is why having participation at all levels of education is so important to creating diversity in STEM, as they show different (but equally crucial) ways to attract and foster underrepresented groups into the field. These programs involve K-12 students in fun, hands-on learning experiences by which historically underrepresented students discover an interest in pursuing STEM degrees and careers. In the latter case, one may be able to facilitate interactive science fairs, establish coding clubs, and

DOI: 10.1201/9781003453581-3

support or start one's own robotics competitions that will keep younger children engaged in participating in hands-on experiences beyond the classroom. By making STEM subjects more accessible and by breaking down stereotypes, these programs help mitigate historical norms that have determined who can dominate the fields of STEM ("The Importance of STEM Education for K-12 Students in Low-Income School Districts," 2023).

For K-12, early exposure to STEM at the foundation builds long-term interest. STEM principles also take new forms in other subjects, such as incorporating statistics into social studies or learning about physics concepts within physical education. Students are able to relate STEM across the curriculum by understanding that those same topics also apply daily. By implementing project-based learning, teachers can promote real-world problem solving for their students.

In college programs, the resources help people get ready for work. A big driver is scholarships, particularly those aimed at underrepresented students in STEM. This type of financial outreach makes it easier for individuals to seek higher education and therefore breaks the chains of economic barriers. This is as important as mentorship programs at the college level. They also connect students with seasoned professionals who provide advice, information, and validation. The mentorship not only provides a need-based curriculum with case studies from the industry but also gives students precious information about what STEM life is really like and empowers them to fit in better in the academic as well as professional worlds.

Research opportunities at the college level further prepare students for future STEM careers. Students participating in research projects have a chance to realize theoretical knowledge of disciplines to be implemented when addressing real-life problems, and to then form critical thinking and generate new ideas. College initiatives usually involve partnerships with industries and labs that offer the students internships and practical experience. Crucial to blending academic studies and the work world, these experiences help students make a seamless transition into technical jobs. These experiences are key to bridging the gap between academics and the job market, ensuring that graduates are well equipped for technical roles. Moreover, active learning approaches adopted by many colleges focus on interactive and inquiry-based teaching methods rather than traditional lectures, thereby engaging students from all backgrounds (Romney & Grosovsky, 2021).

Another way to promote diversity in STEM is through vocational training, something that might attract students who prefer practical, application-based education. These programs are oriented toward jobs and skills that can be employed immediately. This makes vocational training an attractive option for some students who learn by doing as opposed to via theory. Many engineering programs partner with local industry to ensure that coursework is aligned with current job market needs, enhancing work opportunities after receiving a degree. Last but not the least are apprenticeships. This work-study arrangement enables students to learn occupational skills while attending classes and getting paid, which in turn decreases the need for loans.

3.3 STEM Initiatives Overview

Initiatives, ranging from policy exploration to school and community programs, shed light on how the future of STEM education is being formed in various contexts across the world. Between kindergarten through college and beyond, these preparations are administered at disparate levels and each program comes with its own unique personas, approaches, and methodologies. In addition to showcasing what is working, evaluation of these programs also reveals challenges and areas for improvement. This is critical to sense the effectiveness of these programs, and their scalability potential, with the impact on students' development in critical thinking, problem-solving skills, and career readiness.

3.3.1 K-12 STEM Programs Overview

Widely accepted as laying some of the first bricks of education, K-12 STEM is an incubator for curiosity and critical thinking skills in children. There are indicators identifying that early exposure to STEM disciplines has a major impact on students' intellectual development and their future career choices. Kids who interact with STEM early in life establish essential academic skills while fostering an inquisitive mindset that sparks a passion for lifelong learning.

One of the primary benefits of early STEM education is the promotion of curiosity. Children have an innate desire to explore their environment, and incorporating STEM subjects at an elementary level capitalizes on this natural inclination. By presenting scientific concepts in an accessible and engaging manner, educators can spark a lasting interest in these fields. For instance, inquiry-based activities that require students to experiment, hypothesize, and analyze outcomes nurture a deep sense of wonder and exploration.

Combining hands-on projects and technology into the curriculum is another important element of successful STEM education in K-12 settings. Active learning opportunities where students manipulate substances or tools in the physical world or through technology to solve challenges greatly increase student engagement and understanding. These opportunities enable students to relate theoretical learning to real-life scenarios, consolidating their understanding. Project-based learning, such as building simple machines, coding basic programs, or conducting environmental studies, offers meaningful contexts to ground complex concepts in real-world scenarios.

Collaborative opportunities with STEM industries and in schools help create more engaging lessons where students immediately see the value of what they are learning in the classroom. Collaboration with local businesses, universities, and research institutions creates opportunities for students to engage with professionals working in STEM fields. It opens students' eyes to the diverse ways they can explore a career and builds their awareness of the real-world application of STEM knowledge beyond school. Examples include job shadowing, mentorships, and field trips to industrial sites that showcase uses of mathematical and scientific principles in practice.

The availability of teacher training and resources is critical for the effective delivery of STEM content at the K-12 level. Choosing professional development opportunities that enable educators to share best practices as their understanding of high-quality educational methods and approaches to content knowledge expands is an effective way to ensure students do not miss out on STEM skills training that will guide their future studies. To that end, professional development offerings that center on innovative pedagogical strategies, emergent educational technics, and cross-disciplinary routes are needed. Having access to high-quality resources—including up-to-date laboratory equipment and digital devices—allows teachers to create stimulating learning environments that engage and stretch student thinking.

According to "K-12 STEM Education for the Future Workforce" (n.d.), investments by federal agencies, state governments, nonprofits, and industry leaders have positioned STEM education as a national priority. Targeting early childhood education, these efforts aim to build a strong foundation for future workforce development, underscoring the importance of equipping all students with critical thinking and problem-solving skills necessary for high-demand careers. To scale effective models and address challenges like rural underreach, dedicated resources and research interventions remain critical.

Guidelines integrated into K-12 STEM programs provide continuity and effectiveness from one educational stage to another. By understanding the importance of setting specific objectives related to student outcomes, employing various pedagogical strategies, and regularly assessing the success of the program through student evaluations and academic performance metrics, we can develop more effective educational offerings. Network and Organize: Not only is networking important to support individuals in obtaining employment, it is also critical to building awareness of programs dedicated to helping underrepresented groups have access to quality education and resources within the STEM and analytical space.

Additionally, emerging technologies in STEM subjects help students to keep up with rapid advancements in fields like AI and information technology. Exposing students to state-of-the-art tools helps them train for technology-rich environments, but beyond this, it trains students to adapt, a vital skill in an ever-developing world. Teaching coding, robotics, and digital literacy as part of lesson plans helps prepare students for future challenges.

3.3.2 Effect of College STEM Programs

Specialty courses are critical in college STEM programs for preparing students to have the technical knowledge required in areas such as engineering, biotechnology, and computer science. One is taught per an industry standard, rigorous curriculum that gives one the theory behind the process as well as the skills needed to connect the two. In engineering, for example, students work on authentic projects that replicate real-world situations in which they can enact their learning to conduct design and innovation tasks. Biotechnology courses, for example, typically involve lab work in which students manipulate various genetic or cellular processes, giving them hands-on experience that is fundamental to the classroom learning used to comprehend the advanced concepts of biological systems. Furthermore, in engineering, coding bootcamps and software development projects are used to teach students languages and frameworks needed in the technology and robotics industries.

Undergraduate research opportunities are another key element of College STEM programs. Conducting research also fosters creativity in students and enables them to think critically, which is a skill relevant to their future career path. In fact, research work enables students to contribute unique findings and solutions to their respective fields that are not yet explored. Research projects challenge students to think both critically and creatively, whether they are working to develop renewable energy technologies or investigating innovative medical therapies. This indirect, several-step method of inquiry and discovery helps promote the exact mindset useful not only in academia but also transferable across work sectors.

Internships and partnerships with STEM industries are vital in bridging the gap between academic theory and hands-on application. This gives students real-world experiences by working alongside businesses and other organizations. Internships can help one gain experience in professional environments and industry practices that are not covered in the classroom. Through these experiences, students learn professional dynamics, industry expectations, and quickly adapt to technological advancements. Industry partnerships allow for valuable idea sharing and engagement between students and businesses, resulting in enriched learning that is deliverable to businesses looking for solutions.

University STEM diversity initiatives seek to improve equity and access in college STEM programs. Considering that groups including women and people of color have been historically underrepresented in STEM, these initiatives aim to build an academic climate that encourages and advances everyone toward careers in STEM. Programs may provide mentorship to underrepresented minorities, scholarships, or workshops targeting particular barriers faced by these students. Diversity in college STEM programs is important, as it was recruited to promote diverse thoughts and ideas needed for progress and innovation in any STEM field. In addition, fostering a diverse set of individuals in the STEM workforce is critical to ensuring that solutions are broadly applicable and just in a range of communities.

Students interested in STEM programs should explore guidance on how to best pursue those opportunities. They should now actively pursue which institutions are offering strong specialized coursework catering to their professional interests Also, they should ask if or what research opportunities are available for undergraduates, as such experiences can enhance their education and better equip them for what lies ahead. Internships can also help students get a head start in their careers, applying what they learn in class to practical scenarios and meeting contacts in their chosen field. And finally, for those from underrepresented backgrounds, trying to find programs with good diversity support is important so that one can have access to resources that will help one succeed.

3.3.3 Workforce Skills and Vocational STEM Programs

Vocational STEM programs offer a hands-on approach that directly addresses industry needs and skills gaps by equipping individuals with the technical skills necessary for high-demand fields such as robotics and information technology. They are less about imparting broad knowledge than teaching students specific skills, which is why their curricula are developed in collaboration with industry experts. This is so that graduates will have each of the competencies needed to carry out these specialties as soon as they join the workforce.

A hands-on approach is the most effective way to prepare students for careers in STEM, as it helps to connect theoretical knowledge with real-world experience— both of which are vital components of any educational path. Students work on hands-on projects that replicate real-world workplaces to help design their problem-solving training and technical skills. For instance, a student specializing in information technology might work on network configuration or cybersecurity measures as part of their curriculum. This emphasis on experiential learning not only enhances understanding but also builds confidence, enabling students to meet job market demands effectively. Graduates often leave these programs with certifications that validate their expertise, making them attractive candidates to employers seeking ready-to-work professionals.

In addition, one of the most significant advantages of vocational STEM programs is their flexible curriculum design, easily adapted to the rapid pace of technological advancements. As industries develop, so do the abilities needed from their employees. As one expert put it, "the programs have no lag time in applying emerging technologies or changing training mechanisms." The advent of more widely used advanced manufacturing technologies like three-dimensional (3D) printing, for example, allowed vocational programs to ramp up related modules quickly. Such adjustments enable students to stay ahead of the curve in the face of technological advancements and future challenges.

Partnerships between vocational institutions and local businesses further enhance the alignment of educational outcomes with industry expectations. These collaborations facilitate curriculum development, ensuring that training programs are both relevant and comprehensive. Local companies often provide insights into current industry practices and future skill requirements, informing the direction of vocational training. These partnerships frequently lead to internship opportunities and direct job placements, giving students a head start in their careers. An apprenticeship at a leading technology firm, for example, allows students to apply classroom-learned concepts in a professional setting while building valuable networks within the industry

Community colleges are at the forefront of implementation of inclusive vocational STEM programs that help diverse populations. They provide low-cost education options and commonly establish articulation agreements with four-year schools, facilitating student transfers for those desiring additional education. Programs like Trade Adjustment Assistance (TAA) and Career and Technical Education (CTE) grants help support these programs by providing funding for the equipment, tools, and resources needed to ensure quality training. These efforts highlight the role of vocational education in preparing the workforce and its potential to stimulate economic growth through bridging skill gaps.

The demand for continual skill upgrading is another critical factor addressed by vocational STEM programs. Vocational institutions often provide pathways for lifelong learning, encouraging graduates to return for skill refreshers or additional certifications as industries evolve. This commitment to ongoing education helps maintain workforce relevance and supports career advancement in rapidly changing job markets.

By aligning their curricula with industry needs and incorporating cutting-edge teaching methods, vocational STEM programs play an instrumental role in preparing a technically proficient workforce that can sustain economic resilience and competitiveness. For instance, the development of green skills prepares graduates for careers in renewable energy and sustainable manufacturing sectors, which becomes crucial as global priorities slowly shift into the direction of environmental sustainability.

The importance of PPPs in fostering vocational education cannot be underestimated, as they harness the resources and expertise of government bodies, educational institutions, and private companies. One notable example of a successful PPP is Pathways in Technology Early College High Schools (P-TECH), where students graduate with both academic credentials and work experience that greatly enhances employment opportunities.

3.3.3.1 Comparative Analysis of Program Effectiveness

To gauge the success of STEM programs at various educational levels necessitates analysis of multiple metrics, such as graduation rates, employment statistics, and student satisfaction. These indicators reflect the success of STEM initiatives, as well as serve as benchmarks to compare against. For example, in K-12 programs a high graduation rate usually indicates effective engagement and that students are appropriately prepared to enter the next stage of the education continuum. Conversely, college STEM program graduations more accurately reflect that students have obtained the technical skills they need to pursue careers.

Program effectiveness is also evaluated through student satisfaction. Higher retention rates and better learning outcomes are often synonymous with student satisfaction, demonstrating a program's ability to support students with educational and career goals. In on-the-job STEM programs, which emphasize short-term industry requirements, satisfaction could be related to the alignment between the curriculum and the job market and the presence of such fields of expertise.

Examining socioeconomic factors is vital in assessing the reach and accessibility of STEM education. K-12 STEM programs often encounter disparities based on region or community wealth, impacting resources available to schools and subsequently affecting student performance. College programs sometimes have financial barriers that can deter potential students from low-income backgrounds and hinder diversity within these fields. Vocational programs may offer a more accessible pathway due to shorter duration and lower costs, providing opportunities for individuals from varying socioeconomic backgrounds to gain valuable skills quickly.

Tracking advancement in STEM programs requires longitudinal studies. Such studies provide an understanding of the long-term benefits of STEM education by tracking the pathways graduates take in their careers, looking at job placement rates, and tracking career advancement. For example, data might indicate that graduates of certain college programs take on leadership roles in STEM industries more often than graduates of other college programs, suggesting that those programs are effective in preparing students for higher level roles. Similarly, vocational program alumni might exhibit high job stability and satisfaction due to direct entry into the workforce with specialized skills.

We see the challenges in each of these program types reflect the opportunities for improvement and innovation. K-12 programming often struggles with funding for advanced technology and hands-on learning experiences. Forming partnerships with technology companies or securing government grants could further help address these issues, as a wider array of resources would lead to better learning opportunities. One big part of the challenge for colleges is how to keep curricula relevant in an environment that is changing so quickly. The constant need to update and revise courses to keep pace with the changing industry is a challenge in itself. Vocational programs face the challenge of overcoming the perception that they are second-class alternatives to college degrees. Such programs can raise their prestige by displaying success stories of their alumni and stressing how graduates are ready for immediate employment.

In fact, we find that every educational tier offers its own unique contribution to providing STEM education and workforce readiness. Understanding their unique impacts allows educators and policymakers to tailor improvements that extract that which is weakest while enhancing what is strongest. Reform in K-12 STEM education may therefore involve optimizing the availability of resources and trained educators, and ensuring early exposure with gradual engagement over time. College programs that can better align their coursework with industry needs and broaden research opportunities have the potential to enhance student preparedness for complex technological careers. The good news is that making STEM education available and accessible across every level of the continuum is going to broaden opportunity, fuel innovation, and ensure global competitiveness in science and technology.

3.3.4 The Future of STEM Education

In today's rapidly evolving educational landscape, one of the most exciting areas of development is the integration of AI and machine learning into STEM curricula. These technologies promise to create new avenues for learning and discovery, offering students opportunities to engage with material in innovative ways. AI can personalize education by adapting to individual learning styles, providing real-time feedback, and predicting future learning paths based on past performance. This level of customization not only enhances students' understanding but also increases motivation by catering to their personal interests and capabilities. According to Xu and Ouyang (2022), leveraging AI allows educators to streamline instructional design and improve pedagogical approaches, fundamentally transforming how STEM subjects are taught.

Another important trend in STEM education is the rise of online and blended learning models, enabling access to quality education on a global scale as more institutions begin to embrace this delivery method. These models are flexible and allow learners from different backgrounds and geographic locations to access STEM education. Innovation in education can be connecting people through a platform where all learners can learn or share what they have learned and can connect with each other, which will result in sharing knowledge with one another. Through such platforms, extensive conferences and combined classes like online and multiple classes can also be held in different parts of the world on various subjects, enabling a better approach to learning. In fact, Zhang and Aslan (2021) note that the availability of such models potentially democratizes education, providing students from less privileged areas access to the skills essential in a technology-driven milieu.

An increasing focus on interdisciplinary approaches and collaboration has made STEM skills applicable to a wide range of disciplines. Educational programs that integrate STEM disciplines with subjects such as art and humanities encourage creativity and critical thought, both qualities that are becoming more valued in the contemporary job market. Such cross-disciplinary initiatives break down traditional silos in education, creating spaces for students to observe how STEM converges with other aspects of daily living. These approaches not only prepare students as specialists, they also support their development as globally competitive individuals able to solve complex, real-world challenges by working in teams across disciplines.

In addition to industry engagement and providing support mechanisms, policies and funding initiatives are also pivotal to the sustainable and ongoing development of these STEM programs. Such programs need investment by governments and educational institutions—investment to keep them up to date, relevant, and impactful. Government policies that support teacher training, curriculum development, and technology acquisition to meet STEM education needs are crucial for creating a strong STEM education system. Funding from the public and private sectors can also fuel innovation: it ensures that resources are available to implement cutting-edge teaching methods, relevant technology infrastructure, and staff training. These materials, if appropriately allocated, can facilitate improvements in learning outcomes, as they allow for schools to evolve alongside advancing technologies and industries that increasingly depend on STEM talent.

With these trends on the rise, it is clear that STEM education has a bright future ahead, but one that will need to be nurtured and cultivated. Nonetheless, educators need to be proactive, embracing new technologies but also staying cognizant of challenges presented by differences in access and resources. In addition, in order for teachers to skillfully integrate AI and interdisciplinary ways of teaching into their teaching styles, they must constantly develop within their profession. Likewise, stakeholders need to jointly utilize effective design principles for the creation of inclusive and equitable STEM programs that address the diverse needs of learners.

The ultimate objective of all the initiatives is to prepare the students for a workforce where STEM skills are becoming more significant in addressing the challenges of the future. With the ever-evolving landscape of technology impacting various sectors and society as a whole, education systems must transform to equip students with the knowledge and skills necessary to make a positive impact in their respective communities and the world at large.

3.4 Notable STEM Programs

STEM programs at primary, secondary, and higher education levels are able to foster and inspire the next generation of our technology-focused workforce. By examining practical examples and methods, we learn how these programs succeed in engaging students and preparing them for careers in STEM. An exploration not only emphasizes examples of the innovative practices in place but showcases why education must adapt to the needs present in industries today.

3.4.1 Project Lead the Way

Project Lead the Way (PLTW) represents a transformative approach to STEM education, designed to engage students by offering a dynamic curriculum grounded in hands-on, project-based learning. This approach provides students with an opportunity to learn through active participation, engaging them in real-world scenarios that make learning both relevant and exciting. The impact of this teaching method is evidenced by a study indicating PLTW students outperform their peers in school readiness and interest in STEM careers (Project Lead the Way, 2016).

At the heart of PLTW is a focus on experiential learning. There are projects that simulate real-world problems being solved in STEM fields, so students aren't just learning, they are involved in the process. It is through this mode of experiential learning that they realize how the theoretical knowledge read in books can actually be applied, helping them retain concepts better. For example, students will experience state-of-the-art tools and technologies similar to those seen in professional environments, which allows for exposure to various career paths (Access Our Approach, Impact & Efficacy | about Us | PLTW, n.d.).

Another important aspect of PLTW is its dedication to professional development for teachers. What sets PLTW apart from typical programs is the investment in teacher training that is done through the signature Core Training, which is their most collaborative training and is delivered by PLTW Master Teachers who are certified and currently collegial practitioners in the field. One can learn using new data up to October 2023.

In addition, PLTW promotes a collaborative environment for students. Through group projects, students learn to collaborate together to solve challenges, which helps develop vital qualities such as communication and teamwork, and even leadership skills. This kind of collaborative environment is in line with the modern day workplace where one has to work in teams. These team projects help students value diverse perspectives and learn from each other's strengths and prepare them for future careers in a wide variety of fields.

Importantly, the PLTW curriculum is designed with a focus on aligning with real-world career opportunities in STEM fields. This alignment ensures that the skills and knowledge gained are directly applicable in professional settings. Courses are structured to expose students to various STEM careers, helping them identify potential pathways and motivating them to pursue further education or training. The curriculum's relevance to contemporary job markets makes it particularly effective in maintaining student interest and engagement, as evidenced by the high percentage of students who express increased interest in STEM after participating in PLTW courses (Access Our Approach, Impact & Efficacy | about Us | PLTW, n.d.).

As part of its curricular framework, PLTW implements an activity-, project-, and problem-based (APB) instructional model that empowers students to drive their own learning. This model entails a sequence of scaffolded exercises of increasing complexity, which simultaneously enable students to acquire independence when tackling open-ended design problems. Confidence also builds through the process as students work through challenges, applying concepts they have learned to new situations, fostering critical thinking and problem-solving skills.

Real-world applicability is integral to PLTW's approach. For instance, in biomedical science courses, students engage in projects that require them to simulate medical diagnostic processes, gaining insight into the healthcare sector. This exposure not only enhances learning but also provides clarity on the multifaceted nature of STEM professions. Kelsie O'Brien, a PLTW Biomedical Science alumna, articulates how such experiences contribute significantly to personal and professional growth, underscoring the program's value.

Ultimately, PLTW prepares students with transferable skills necessary in many different fields, including critical thinking, collaboration, and moral reasoning. These are skills increasingly in demand by employers, hence help to build a strong foundation for academic achievement and success in the workplace. PLTW integrates these skills into the learning process so that students can thrive in an ever-evolving world, becoming well-rounded individuals who are prepared to face future challenges (Project Lead the Way, 2016).

3.4.2 STEP Initiative at the Florida Consortium

The STEP initiative, formally known as the Florida Consortium of Metropolitan Research Universities STEM Talent Expansion Program (STEP), serves as an impactful model for improving retention and graduation rates within STEM disciplines. By examining the goals and outcomes of this program, we aim to understand how targeted interventions and support structures can effectively nurture underrepresented students in STEM pathways.

At its core, the STEP initiative is driven by a critical goal: to increase retention and graduation rates among students pursuing degrees in STEM. Retention rates remain a significant concern within STEM fields, with many students abandoning these majors due to various academic and social challenges. The STEP initiative addresses these issues by providing a structured environment where academic success is fostered through personalized support and resource allocation (NSF Scholarships in Science, Technology, Engineering, and Mathematics Program [S-STEM], 2024).

Among the core strategies of STEP are financial support and resources for underrepresented students. These scholarships are not just allowances—everything is a lifeline program for students struggling to find educational financing. Reducing this financial burden can also enable students from disadvantaged backgrounds to concentrate even more on their studies, without worrying so much about their economic welfare. Which means from a financial aspect, with the STEP, financial stability is never a barrier for student academic persistence.

Additionally, this program works closely with companies to offer internships. Internships are the bridge between academic theory and practical hands-on application. They provide students with practical application that enriches their education and paves the way to a career. By collaborating with corporations and institutes, STEP creates direct pathways for students to engage with professionals from their respective fields, thereby enhancing their educational experiences with invaluable insights and real-world experiences.

Evidence-based interventions form another key component of the STEP initiative's methodology. These interventions are meticulously designed to tackle specific academic challenges that students encounter during their STEM education. For instance, workshops and tutoring sessions focus on difficult subjects, ensuring that students have access to the guidance necessary to overcome obstacles in their coursework. Such targeted interventions are based on extensive research, aligning with proven strategies that promote student success and retention.

STEP is a collaborative endeavor and brings together faculty, industry experts, and mentors who work together to establish a network of support for students. This network is essential, as it provides students with mentorship and encouragement, which are crucial factors in maintaining motivation and persistence in demanding STEM programs. The program helps students cultivate resilience against pressures and challenges common among STEM students by creating a community and sense of belonging.

Moreover, the STEP focuses on exposing students to the consumption of STEM content. Not only do these activities help reinforce learning, they also deepen students' ties with their academic communities. Participating in such initiatives provides students with opportunities to test classroom learning in real-world settings, which strengthens their commitment and interests in their respective fields (Shortlidge et al., 2024).

The STEP initiative's outcomes show that it is delivering on its goals. Participants in the program have reported higher retention and graduation rates, signifying that the program is fostering a supportive environment in which participants can thrive academically over the long term. This initiative provides an opportune means to address psychosocial as well as academic needs that lead to the confidence and skills necessary for students to thrive in STEM-related fields.

Guidelines for implementation of similar programs stress the importance of tailoring interventions to the specific needs of participating students. Identifying student diversity necessitates considering variables like socioeconomic restraint, unique cultures, and academic challenges. This approach involves ongoing evaluation and adjustment of various program elements, ensuring that they stay relevant and effective in meeting the changing needs of students.

3.4.3 Louis Stokes Alliances for Minority Participation

The Louis Stokes Alliances for Minority Participation (LSAMP) program addresses the underrepresentation of minority groups in STEM education and employment through strategic partnerships with educational institutions, research organizations, and industry stakeholders. Founded in 1991, LSAMP aims to foster a more inclusive environment through the advancement of underrepresented individuals across a wide range of educational pipelines in STEM (Louis Stokes Alliances for Minority Participation, 2024). This model is based on collaborations among institutions to create an ecosystem in which minority students can achieve academic and professional success in STEM fields.

The LSAMP program is an NSF initiative that is primarily designed to increase minority participation in STEM educational pathways. This entails engaging in partnerships with colleges and universities to establish strategies that successfully recruit and retain students who have been historically underrepresented. These groups include African Americans, Hispanic Americans, American Indians, Alaska Natives, Native Hawaiians, and Native Pacific Islanders. Through collaboration, LSAMP employs evidence-based strategies to develop environments that nurture the learning and success of these students. In doing so, LSAMP positions its mission with national interests focused on growing the preparation of the STEM workforce for future technological advancements and innovations (Louis Stokes Alliances for Minority Participation [LSAMP], 2015).

The core of LSAMP's initiatives is academic support and mentorship. By offering personalized guidance to minority students, the program plays an integral role in reducing barriers to STEM education. LSAMP cultivates student–faculty mentoring bonds creating the necessary support that student's need to navigate their academic careers, encouragement, and access to resources. Mentorship Programs are aimed at developing students' confidence, academic skills, and provide brief insight into career pathways in STEM. These mentor–mentee relationships often transcend the academy and provide students support networks that help advance them within their professional journeys (Louis Stokes Alliances for Minority Participation [LSAMP], 2015).

LSAMP also hosts workshops and conferences to help minority students build critical leadership skills, in addition to offering mentorship. These activities are centered around skills like problem-solving, teamwork, and communication—invaluable skills not only in academia but the workforce as well. Often, workshops are led by experts in the STEM fields, who share their journeys and offer firsthand insight into how to thrive in the many different types of STEM careers available. Conferences provide opportunities for students to showcase and discuss their research, broaden their professional networks, and develop confidence in public speaking and presentation skills (Louis Stokes Alliances for Minority Participation, 2024).

It is important to measure the success of LSAMP initiatives to make sure that they are effective. Success is measured mainly by how many people of color earn STEM degrees. Using graduation rates and degree completions as metrics, LSAMP tracks how effective its strategies are in promoting student success. LSAMP has helped cultivate a diverse workforce of future STEM professionals, as reflected in the higher graduation rates of participating students from minority backgrounds, gender, or other underrepresented demographics. Additionally, the program places emphasis on ensuring that these graduates are well-prepared and highly qualified to pursue further education or enter the STEM workforce, thereby contributing to diversifying the nation's scientific community (Louis Stokes Alliances for Minority Participation [LSAMP], 2015).

LSAMP's approach further includes partnering with a range of institutions, from community colleges to research-intensive universities. These relationships help them share information and best practices for recruiting and retaining minority students. For example, community colleges may offer foundational education prior to students migrating to four-year establishments, and research universities might provide each advanced coursework and research alternatives. This collaborative framework allows for shared resources and expertise across institutions, ultimately supporting students along the entire pathway of their STEM education (Louis Stokes Alliances for Minority Participation, 2024).

Moreover, LSAMP's partnership structure places a higher priority on the quality of individual partners than on the total number of partners, assuring that each partner is a strong contributor to the effort to increase the number of underrepresented minorities pursuing STEM fields. The varying organizational partners bring unique strengths to the table. For example, liberal arts colleges may focus on teaching pedagogy, while major research centers could provide research experiences, creating a holistic educational experience for students. This alliance model fosters relationships that amplify the reach and impact of LSAMP's objectives, maximizing opportunities for student engagement and success (Louis Stokes Alliances for Minority Participation [LSAMP], 2015).

In conclusion, the LSAMP program demonstrates a collective approach to championing targeted interventions toward enhancing diversity in STEM. Through its emphasis on minority representation, strong support systems, and concrete metrics for measuring outcomes, LSAMP is leading the charge toward a more inclusive and dynamic STEM workforce. The national effort to provide new innovations and adopt evidence-based practices will be critical in addressing existing disparities and preparing a new generation for the opportunities and challenges ahead in a science- and technology-based world (Louis Stokes Alliances for Minority Participation, 2024).

3.4.4 Northeastern University Cooperative Education Program

The co-op program at Northeastern University is the original real-world work–academic study program, providing student participants with payoffs that pay off in real life. Northeastern's education model is rooted in an immersive learning experience that has been redefining education for decades.

A highlight of this program that makes it unique is that for the first two or so years, each semester alternates between coursework at school every other semester and full-time employment in a job related to one's major. This approach enables students to implement theoretical knowledge acquired from their classes in real-life circumstances within the workplace. Students enrolled in hands-on engineering curricula, for example, may spend one semester learning about complex mathematical models and computer-aided design (CAD) software, and the next applying those skills in the field, such as on projects with companies like SpaceX or at Massachusetts Institute of Technology's Plasma Science and Fusion Center (Northeastern University, 2024).

Not only does this rotation give students the opportunity to develop practical skills, it also enables students to begin building their professional networks, connecting with mentors and industry leaders who care about their success. For example, Danielle Murad Waiss's co-op at NATO Defense College highlights how students can forge relationships not only on campus but also on the world stage, contributing to personal and professional development (Undergraduate Co-Op—Northeastern University College of Engineering, 2024).

In addition, Northeastern's co-op program facilitates students' ability to reflect on what they learned academically and apply it to their work experiences. The program features a preparation stage, where students must take a co-op course. The tools they gain in this course not only aid them in finding jobs but also teach them how to act in the workplace, allowing them to complete co-ops and contribute to them. While and after they are in their placements, students reflect through assignments and dialogues with coordinators, and identify how their experiences translate into learning, relating them to their studies (Undergraduate Co-Op—Northeastern University College of Engineering, 2024).

Such reflective exercises are important as they prompt students to consider their roles and impacts in a professional context. Students gain a deeper understanding and prepare themselves for future challenges by thinking about how their work relates to what they learned in the classroom. This extensive reflective process allows students to engage meaningfully and expensively with professional environments rather than simply being given exposure to professional environments (Northeastern University, 2024).

Being in a co-op program provides a much greater resume-building experience with very meaningful internships. As many students realize, the experience and relationships built in their co-op placements provide a competitive advantage when looking for jobs. Statistically, 58% of Northeastern students receive job offers from former co-op employers, and they earn a starting salary that is 33% higher than the national average. These numbers demonstrate the efficacy of experiential learning in postgraduation career preparation (Northeastern University, 2024). Additionally, co-op students tend to have a seamless integration from education to career. The confidence and skills that they gain from these firsthand experiences are immeasurable and cut down the distance we usually see when transitioning from student to working professional. That journey transforms them and shortens the pathway to being work-ready (Undergraduate Co-Op—Northeastern University College of Engineering, 2024).

But there are wider benefits to community and industry as well. The program fosters a well-prepared graduate who reaps hands-on experience to supply the need for skilled workers to solve contemporary issues and innovate in the workforce. This relationship between academia and industry provides sustainability and relevant industry standards (Northeastern University, 2024).

For its part, Northeastern has been constantly evolving its co-op offerings to accommodate the needs of the wide range of students to whom it serves; one size doesn't fit all. Whether it's helping students land placements locally or overseas or providing stipends for lower-paying roles in nonprofits and startups, the program emphasizes accessibility and inclusivity. Such flexibility showcases Northeastern's commitment to ensuring every student has the chance to engage in powerful experiential learning opportunities, irrespective of their financial background or chosen major (Undergraduate Co-Op—Northeastern University College of Engineering, 2024).

3.5 Advantages of STEM Exposure for Marginalized Groups at an Early Age

In order to justify the importance of introducing marginalized communities to STEM at an early age, it is important to be able to recognize that there are various initiatives that significantly benefit from this type of engagement. In part, those pipelines don't exist because early exposure to careers in STEM can have a significant impact on the academic and professional pathways of underrepresented young people.

For starters, involvement in STEM activities at a young age can develop problem-solving abilities, abstract thinking, and improve analytical skills. And it is not only in the classroom that such cognitive gains are possible, with these skills also transferring over into much broader life experiences that will allow our children to think more critically and creatively about world problems. When children conduct basic experiments, or when they engage in very basic coding exercises, they get an up-close-and-personal understanding of how to form a hypothesis and how to go about testing out those ideas based on what the results appear to be. The amazing thing about this hands-on experience is that it creates a foundation for their critical thinking to develop and become a huge asset in their life.

In addition, engaging parents and caregivers through community workshops contributes to changing cultural norms around STEM. In the process, they provide an opportunity for many families who may feel estranged from STEM due to unfamiliarity or historical exclusion (such as people of color) to demystify these subjects. These workshops create a network to support home learning around STEM by offering fun, accessible activities and strategies. They also engage parents and caretakers to actively participate in their children's learning, creating programs where the curiosity for STEM is embedded and appreciated.

Trends indicate higher retention rates and sustained interest in STEM disciplines for those who are exposed at an early age. Programs such as after-school robotics clubs or science camps create a sense of community among students and provide them with a platform to explore their interests in depth. When students see the practical applications of what they learn, they're more likely to stay engaged. This increased engagement often translates to better performance in school and a stronger likelihood of pursuing advanced studies in STEM fields later on.

The advantages of early STEM exposure are hardly unique to the student's own academic achievement. The importance of early involvement in STEM education is strongly associated with long-term career advancement, economic independence, and wider social benefits. For example, a young girl who develops a passion for engineering through a school robotics program might go on to earn a degree in the field, eventually securing a well-paying job. This will not only improve her financial stability but helps to break the cycle of poverty in which she lives. Even better, she starts to serve as a role model and inspire girls in younger generations to follow suit, eventually shifting the demographic in STEM fields.

In addition to individual success stories, increasing diversity in STEM brings fresh perspectives and innovative ideas to the table. Diverse teams in STEM fields are better equipped to tackle complex societal challenges, such as climate change, healthcare disparities, and technological advancements. A workforce that reflects a variety of backgrounds and experiences leads to more comprehensive solutions that address the needs of all communities. Thus, promoting early STEM engagement among marginalized groups is not just an educational imperative but a societal one.

Requesting policy support at multiple levels of education to keep these efforts running is crucial. Funders might begin supporting programs to get the ball rolling on early STEM education in underserved communities. Whether it is grants or scholarships, students can benefit from a little helping hand to make sure they get the resources and opportunities that will help them in their education. A dedication to the professional development of STEM teachers further enhances their capacity to provide instruction.

It's equally important to weave STEM concepts throughout every class, not just relegate them to particular science and math classes. When educators interlace STEM with history, art, and physical education (PE), they are depicting how STEM application can enhance the scope of other subjects. For example, incorporating physics of movement in a PE class or utilizing statistics within social studies can pique the interest of students who are not inclined to traditional STEM courses.

A proven strategy of increasing engagement and relevance in STEM subjects is incorporating project-based learning. This could involve real-world, applied work like "getting students outside to focus on environmental cleanup," or giving students the ability to design machines that solve simple problems and then build them, through a class where students would actually create their websites and learn (at least some) of the code behind those sites. Furthermore, most of those projects are collaborative and help students to develop their skills in teamwork and communication.

The best after-school and summer programs add another layer of exposure to STEM content that might not be part of the regular school day. Coding clubs, robotics competitions, science fairs allow students to explore their interests further and innovate on new ideas. These programs also develop a sense of achievement and community among members, something that may encourage persistence in STEM education (and careers) over the long haul.

Again, mentorship programs are also an invaluable way to guide students, while giving real-world perspective to careers in STEM. I believe students also benefit from having a mentor who can offer critically important advice and support, and expose them to STEM career paths known to them. Mentors provide living examples of career success in STEM, reaffirming to students that anyone is capable of thriving in such fields.

Lastly, addressing challenges of accessibility in STEM education means that all students, also those who are disabled, have the same opportunities to be success-ful. Our schools are to promote an inclusive environment through accommodation, which should include the following: American Sign Language (ASL) interpretation, wheelchair accessible venues, and adaptive technologies requirements. If we democ-ratize STEM education, then every student will have an opportunity to learn and contribute something extraordinary within the field.

3.6 Successful Narratives From Diverse STEM Professionals

Personal stories and journeys of STEM professionals from diverse backgrounds serve as powerful tools to emphasize the effectiveness of inclusion initiatives. These narratives not only highlight the resilience and determination required to succeed in STEM but also provide relatable pathways for younger generations to follow.

One such story is that of Dr. Mae Jemison, the first African American woman to travel into space. Her journey was marked by significant challenges, including over-coming racial and gender biases in a predominantly White and male field. Despite these obstacles, her determination and passion for science led her to achieve her goals and inspire countless young women of color to pursue careers in STEM. Stories like hers demonstrate that success is attainable, no matter the barriers faced.

Another inspiring figure is Dr. George Washington Carver, an African American scientist and inventor who revolutionized agriculture in the United States. Born into slavery, Carver faced immense adversity throughout his life. Yet, his innovative work with crop rotation and soil health had a profound impact on farming practices and food production. His story underscores the importance of perseverance and the poten-tial for groundbreaking contributions that can arise from difficult circumstances.

Percy Lavon Julian is a prominent figure in the synthesis of key chemical com-pounds. Racial hurdles necessitated an arduous path to success for Julian, whose work led to the synthesis of physostigmine, a compound administered to treat glau-coma. Subsequently, he developed ways to manufacture synthetic cortisone, lowering the price to a fraction of what it had been and making this crucial drug available to millions of patients with arthritis. His studies laid the groundwork for the factory production of steroid drugs, so important in the treatment of various medical condi-tions today.

Dr. Norbert Rillieux is noted for revolutionizing how sugar was processed. Born to a free Creole family in New Orleans, Rillieux invented the multiple-effect evapo-rator system, which transformed the sugar refining process. Commonly known as the father of the steam power revolution, his invention harnessed steam power more effectively, increasing the safety of production exponentially while lowering costs and enhancing output quality. Not only did Rillieux's system process sugar more effi-ciently, it also addressed hazardous labor conditions related to the older processes, thus proving to be a marvel of engineering whose benefits extended beyond the industry and demonstrated the impact of innovative engineering solutions on society. His legacy is a reminder of the way scientific ingenuity can raise industry standards and improve human welfare.

Another ground-breaking scientist who made a significant impact on medicine was Dr. Alice Augusta Ball. She was a pioneer in early 20th-century academia, building an illustrious career in chemical research as an African American woman. At the University of Hawaii, she created the "Ball Method," the first effective treatment for leprosy, using injectable chaulmoogra oil extracts. Ball's approach provided hope and a cure to thousands suffering from this crippling illness, proving to be a breakthrough in treatment methods. Sadly, she passed away at only 24 years of age, yet this groundbreaking achievement has since gone on to save the lives of many others and bring attention to the need for more minority scientists in the field of public health

Dr. Ellen Ochoa is another pioneer in the realm of space exploration, as she broke the mold when she became the first Hispanic woman astronaut. Her work for the National Aeronautics and Space Administration (NASA) includes some notable milestones, such as the directorship of the Johnson Space Center. Prior to being appointed to this leadership position, Dr. Ochoa completed four space shuttle missions and spent nearly 1,000 hours in space. As the director of the Johnson Space Center, she worked to innovate and improve human spaceflight.

Dr. Patricia Bath invented the Laserphaco Probe used to treat cataracts and revolutionized ophthalmology. This new surgical device enabled more minimally invasive surgeries, which improved patient outcomes greatly, restoring sight to many patients who would otherwise go blind. Dr. Bath's focus on making healthcare more accessible resulted in community ophthalmology, a new specialty combining public health and clinical knowledge to provide eye care to disadvantaged communities. This pioneering work shows how central innovation is to developing new treatments and highlights the importance of addressing disparities in healthcare.

Dr. Charles Drew's pioneering work in blood storage and transfusions has saved millions of lives globally. During World War II, his work established large-scale blood donation and transfusion practices that are so important in modern medicine. The problem was lengthy and expensive, including the short shelf life of stored blood, until his innovations. Dr. Drew's techniques enhanced blood preservation and transportation, resulting in blood being more accessible for use in combat medical situations.

Dr. Chien-Shiung Wu, often referred to as the "First Lady of Physics," conducted pivotal experiments on nuclear beta decay, fundamentally altering the landscape of particle physics. Her work provided critical confirmation of the theory of weak interaction, a fundamental force governing subatomic particles. Known for her meticulous methodology, Wu's experiments in the 1950s disproved the conservation of parity and demonstrated that particles could distinguish between left-handed and right-handed systems. This monumental discovery not only validated theoretical predictions but also showcased the importance of experimental evidence in scientific breakthroughs (Shah et al., 2024).

Dr. Subrahmanyan Chandrasekhar revolutionized understanding of stellar structure and evolution. His theoretical work on the stability of stars was the basis of the formulation of the Chandrasekhar limit, changing the way we think about the life cycle of stars. The Chandrasekhar limit is the maximum mass that a stable white dwarf could have before collapsing under its own weight, leading to a supernova or formation of a neutron star or black hole. This visionary realization set the stage for an entire branch of astrophysical research focused on stellar relics and the processes that alter them, representing one of the most significant advances in the field in nearly a century. Chandrasekhar, known for his exact calculations, won the Nobel Prize in Physics in 1983, solidifying his legacy in stellar dynamics (Stanislav, 2017).

Dr. Wing Ng's research has been instrumental in advancing high-speed aircraft dynamics. His contributions include pioneering techniques for reducing drag and improving fuel efficiency, crucial for the development of modern aerospace technology. By enhancing our understanding of airflow behavior around high-speed aircraft, Ng's work has facilitated innovations in both military and commercial aviation sectors. His developments in turbulent flow control and heat transfer efficiency continue to influence design strategies for next-generation aircraft, ensuring safer and more efficient air travel.

Lastly, Dr. Michael E. Brown's discovery of Eris, a distant celestial body in the Kuiper Belt, played a pivotal role in the reclassification of Pluto as a dwarf planet. In 2005, Brown's team identified Eris, which appeared larger than Pluto, prompting astronomers to reconsider the criteria defining a planet. This discovery led to the International Astronomical Union's decision to refine the classification system, ultimately resulting in Pluto's relegation from full planet status. Brown's findings underscored the dynamic nature of astronomical categorization and highlighted the importance of continuous exploration in expanding our solar system's boundaries.

Networking is also important. Organizations like the Women in Engineering Pro Active Network (WEPAN) provide spaces for women to make connections, learn from their experiences, and grow together. Building community is critical to the retention of diverse talent in STEM, and these networks often provide this sense of community and belonging. In addition, partnerships established between academic institutions and industry sponsors can translate into internships and employment experience for professional development.

STEM provides representation that, in turn, has measurable effects on perceived belonging and motivation. Seeing others like oneself in STEM fields reinforces the belief that one can be successful there. Role models can come from all walks of life—people in a variety of disciplines who reveal that STEM is not reserved for one particular group and whose experiences question stereotypes. Some programs, for instance the "If/Then" Ambassador Program that profiles women who are making a mark in STEM, work to alter how girls see careers by exposing them to successful female role models.

Representation, of course, goes beyond just what we see with our eyes. This includes making sure that voices representative of the diversity of humanity are heard in conversations about what to do within STEM and around STEM policies. This wider inclusivity means more perspectives, the unsung needs and harder truths are heard and met—so more universal solutions will hold water.

Through storytelling as a mechanism for change, scientists who have navigated the path from underrepresented places in STEM should have their turn at policy advocacy and outreach initiatives. What does one gain with narratives? The humanity behind numbers and experiencing what it is like to be a part of a marginalized community. Professionals share their stories to shape public opinion and call for policies in favor of D&I.

Take theoretical physicist Dr. Jim Gates, for example, who often discusses diversity in STEM on his platform. He writes about his own experiences and speaks out for education policy that provides equal access for all students, regardless of their background. His participation in policy talks further demonstrates the need for systemic change to foster a more equitable STEM setting.

Personal narratives are also a great way to engage through outreach initiatives. Organizations like Black Girls CODE, which provides girls of color between the ages of 7 and 17 opportunities to learn computer programming skills, use Women In Tech stories to motivate and mentor the Female Pipeline. Apart from technical skills, these initiatives do much to instill a feeling of confidence and optimism about future prospects.

3.7 Initiatives Taken by the Organization to Promote Inclusive Workplace

Having an environment within STEM organizations that is supportive and inclusive is crucial in both attracting as well as retaining diverse talent. At their most basic level, inclusive policies are policies that recognize the issue and contain steps to counteract those unique disadvantages faced by marginalized groups. Not only do these policies demonstrate an in-depth comprehension of those difficulties, they also indicate that the company prioritizes promoting D&I. For instance, enacting antidiscrimination policies and enforcing equal employment opportunities can help ensure that anyone, irrespective of his or her background, has a fair chance at success.

Organizations have a willingness to offer training programs on educating the staff regarding unconscious biases, and they are keen enough to take leadership initiatives in developing diverse talent. At its core, unconscious bias training is aimed at raising employees' awareness to be able to identify hidden biases that might influence how they take decisions in the workplace. For instance, the Society for Advancement of Chicanos/Hispanics and Native Americans in Science (SACNAS) provides training opportunities that raise awareness and help educate staff so they may treat employees more equitably (McGee, 2021). They said leadership programs that include mentoring and sponsorship of underrepresented groups will further help them with their career advancement.

Affinity groups and regular feedback mechanisms can be powerful to retain diverse talent and adapt the business to what employees need. Affinity groups (some organizations referred to these as employee resource groups) offer an area to network, talk through experiences, and grow within ones culture. Such groups can be influential for promoting changes within the company and providing a sense of community among its members. Employees are able to give feedback and voice their concerns on a regular basis through surveys, town hall meetings, etc., which allows the organization to take corrective measures in order to ensure they have a healthy, happy and balanced life at work.

Companies also partner with industry, facilitating internships and a pipeline to diverse talent from school to work. Partnerships already exist to enable students gain practical experience and network with companies. Programs like the Alaska Native Science and Engineering Program (ANSEP) focus on creating pathways for Alaska Native students from kindergarten through university to pursue STEM careers. These collaborations often include mentorship, scholarships, and internships that help students gain practical skills and exposure to potential employers.

Organizations can indicate how seriously they take diversity by directly confronting the issues that marginalized groups face with more inclusive policies. For example, strong flexible work arrangements and the availability of comprehensive parental leave policies matter a great deal to women and other underrepresented groups. Transparent promotion programs and pay equity policies, too, can help establish fair playing fields for growth and financial security for all employees.

There is no underestimating the value of quality training programs. It is important to mandate unconscious bias training with regular updates as societal norms and expectations change. Leadership development programs need to identify and cultivate diverse talent serving as a direct pipeline of those who will ultimately lead the company, ensuring that our leadership benches reflect the diversity we aim for across our organizations. This type of representation at the very top is instrumental in encouraging others and demonstrates what is possible here (How to Retain Diverse Talent in STEM | PA Consulting, 2024).

Another important consideration is stakeholder collaboration. Working with other organizations or even industries can help make diversity initiatives broader and more impactful. For instance, relationships with groups such as Latinas in STEM can offer Hispanic women looking into STEM careers additional resources and support. Additionally, these collaborations can result in the creation of shared programs aimed at specific populations looking for unique support and skills that may not exist within the organization independently. Diversity and inclusion initiatives need to be assessed on an ongoing basis. One of the ways is by looking at metrics like how many employees one has, diversity among one's leadership, and employee satisfaction scores to determine if these efforts are effective or not.

Furthermore, with the aim of creating an environment in which each and every person feels included, respected, able to bring their best selves to work, organizations can establish the following: Inclusive Policies—Strive to implement inclusive policies that ensure people from all backgrounds feel at home in the organization; Comprehensive Training—Offer comprehensive training programs that ensure all employees are conscious of the importance of inclusivity; Affinity Groups—Create affinity groups for underrepresented minorities and other terminologies alike, where employees have a comfortable circle with whom they can share their experiences; and Industry Partnerships—Collaborate with industry partners who too believe wholeheartedly in increasing diversity within STEM and support each other along the way. Not only is this approach fair and equitable, it is also a source of innovation and competitive advantage for organizations and society in general.

3.8 Final Insights

This chapter mainly focused on different levels of very formal education and their application in diversifying STEM initiative approaches. For the K-12, interactive science fairs and project-based learning activities provide a good way for students coming from different backgrounds to get better exposure and develop an early interest in STEM. Colleges support this initiative by offering scholarships, mentoring programs, and research opportunities to help underrepresented students prepare for STEM professions. STEM accessibility is one of the things vocational training avails, meaning that it gives a practical and application oriented education that reflects the demands of the labor market at any time.

The combined impact of these initiatives is immeasurable. When STEM education is woven across different layers in a community, curiosity and innovation spread like wild fire, which benefits everyone as a whole. These multifaceted efforts serve to promote an atmosphere of inclusiveness while opening up opportunities for underrepresented groups to succeed within STEM fields. It is still essential for educational institutions, industries, and local communities to collaborate in these kinds of efforts. Together, by mobilizing toward a more inclusive and better-prepared STEM workforce, we can find the solutions to an increasingly complex set of challenges.

4

STEM Career Exploration

Economic Benefits

4.1 Comparison of STEM Profitability With Other Industries

Jobs in STEM pay more money and increase long-term economic security and wealth. STEM jobs almost always come with higher starting salaries than the typical non-STEM positions. This leads to better salaries during a work life. Empirical studies have found that, on average, bachelor's holders in STEM earn much more than their counterparts in non-STEM disciplines like the social sciences or arts (Melguizo & Wolniak, 2012). This wage gap is no surprise given the fact that, as a general rule of thumb, STEM skill sets are in such high demand these days and job requirements for these roles can be quite specific.

This economic stability creates a financial advantage that allows the STEM employee to accumulate wealth. Having less money means less to save, invest, and prepare with for the future. This is an even more important consideration for individuals within marginalized sectors. A strong economy means fewer events that can lead to financial disasters and helps cushion one from unexpected happenings. Not to mention, wealth accumulation is what makes possible home ownership, investment in an education, and ultimately, a slightly better life for future generations.

Research on the way that economic downturns affect different industries brings to light that STEM jobs as a whole are more recession-proof. When an economic downturn happens, sectors like technology, healthcare, and engineering tend to do well because their products and services are vital. In fact, the technology sector has been seen to fare better than other sectors and experienced less job-loss during past crises such as the global financial crisis of 2008. The pandemic of 2020 exacerbated this reality; while many other sectors saw deep budget cuts, technology and healthcare industries, both deeply rooted in STEM, kept demand up.

This resilience highlights the importance of STEM careers for long-term job security. Marginalized individuals pursuing STEM paths are therefore likely to benefit from more stable employment even during economic downturns, mitigating the risks associated with unemployment or underemployment that can disproportionately affect these communities.

DOI: 10.1201/9781003453581-4

Examining the educational costs versus potential earnings over a career span also underscores the value of investing in STEM education. While the initial cost of obtaining a STEM degree can be high, the return on investment is substantial. A study by Carnevale et al. (2011) found that individuals with bachelor's degrees in STEM fields have around 84% higher lifetime earnings compared to those with only a high school diploma. These statistics emphasize that the upfront costs of education are outweighed by the long-term financial benefits.

Without access to the higher potential earnings of STEM careers, marginalized individuals pay more for student loans and spend a longer time indebted before achieving financial freedom. This economic independence provides greater financial freedom and the ability to care for oneself, one's family, and one's community.

The promotion of STEM careers among marginalized communities also aligns with efforts to address equity gaps in education and workforce participation. Research has shown that systemic barriers often prevent minority groups from entering and completing STEM programs (Haeger et al., 2024). However, initiatives designed to support these groups through targeted mentorship, scholarships, and inclusive educational practices can make a significant difference.

Educational institutions are very important in this process. Universities can lessen these challenges by fostering D&I. Especially great are programs where students have a real hands-on research experience, internships, and work closely (even collaborate) with industry. These unique face-to-face experiences not only improve skill sets but also cultivate critical professional networks that are essential for the next step in one's career.

Furthermore, systemic changes within educational institutions to promote gender and racial equality can positively impact students' views and persistence in STEM majors. Universities must continue to channel resources toward nurturing the skills and knowledge of marginalized students, ensuring they receive both physical and psychosocial support (Melguizo & Wolniak, 2012).

Over the long term, it turns out that STEM careers are also economically beneficial for more than just the individuals who choose to pursue them. Diversity in STEM workers means more marginalized folks can be part of the workforce, and as a result bring different perspectives that drive innovation. It is important for solving the largest global problems, ensuring that everyone benefits from technological advances.

4.2 Outlook for Minorities in STEM Fields

STEM jobs are changing faster than ever before, opening doors to marginalized individuals in ways that were unimaginable a generation ago. Predicting the future of career fields will be instrumental to the equitable representation of marginalized voices.

A review of hiring trends points to increasing demand for minority candidates in STEM industries. With data science, engineering, and technology, it now turns out that there is a rampant demand for analysts to start showing up in the workforce. The concept that an open global community can drive innovation faster than a closed system is not particularly new, and companies are starting to appreciate this as they see the power of brilliant minds working together. As more and more organizations put D&I strategies into place, minority-targeted recruitment is on the rise.

Upskilling and reskilling have profound impacts on employment prospects for marginalized individuals. For instance, transformative programs, such as those pointed out by the World Economic Forum, lead to lifelong learning, which is built on analytical thinking, creativity, and problem-solving skills (Li, 2022). To promote skill development, online courses, workshops, and certification programs are now used to help people upskill and become relevant for new job roles. As another example, experiential learning in advanced technologies, such as the NSF's ExLENT program, supports transitions and career advancements for nontraditional learners (Education & Workforce Development—Midwest Big Data Hub, 2024).

Partnerships between educational institutions and industry is a major factor in meeting diversity goals in the STEM workforce. Technology companies and research organizations are partnering with universities and colleges to inform pathways for underrepresented students. These partnerships can materialize in the form of internships, scholarships, sponsorships, mentorship, or research. They give students practical experience and a chance to socialize as well, making them more employable. This includes the Minority Serving—Cyberinfrastructure Consortium (MS-CC) that facilitates cyberinfrastructure access for faculty and students at historically underserved institutions (*Education & Workforce Development—Midwest Big Data Hub*, 2024).

An outlook for how these trends will impact the STEM workforce of the future points to a real potential shift toward increased D&I. The push to reflect the audience will likely drive a greater demand for minority professionals. Many of the new technology advancements will require targeted job functions, thus the need to always upskill workers. Projects like AI and machine learning will create new, creative, and off-main-stream jobs which require a larger pool of diverse talent to face the future challenge.

4.3 Realities of Minority Representation in STEM Jobs

The current landscape of minority representation in STEM careers reveals significant gaps that demand our immediate attention. Data show that Black and Hispanic workers remain underrepresented in the STEM workforce compared to others in the overall labor force, including areas like computing that have seen notable growth (Funk et al., 2021). This disparity is evident across various areas, suggesting systemic barriers that many minorities face from education through to employment. By contrasting data from past decades with recent years, we can observe whether minority participation in STEM has improved or regressed. According to federal government data, while there have been some gains, substantial gaps persist, particularly in high-demand fields like computing and engineering (Funk et al., 2021).

Much of this puzzle will be solved in the examination of implicit hiring prejudices and toxic workplace culture. Indeed, per previous studies, biases manifest in different steps of the employment process such as resume screening, interviews, and more surprisingly performance evaluations. Minority candidates have substantially long odds of being hired or promoted when confronted with these biases. Research has actually shown that, with things being equal, minority candidates receive far fewer callbacks in response to their resumes than other job seekers. Next, many minorities are faced with workplace cultures that are either not conducive or outright hostile to their growth into the upper ranks of a profession.

For example, Pew Research Center stated that only around 50% of racial and ethnic minorities who work in science, engineering, and technology jobs feel they are treated fairly in terms of promotions, compared with about 80% of White STEM workers (Funk & Parker, 2018). This means that even if and when minorities do end up in a STEM role, they still face systemic inequalities when it comes to career progression. The only way to paint a clearer picture of this is by conducting comparative analyses and understanding the experiences of different demographic groups within the same industry sectors. For example, comparing the career trajectories of minority women with those of their male and White female counterparts can uncover specific challenges faced by this doubly marginalized group.

Not only that, measuring implicit bias correctly also includes the qualitative culture components of a workplace. Often, feedback from minority employees uncovers subtler but equally insidious forms of bias. These include isolation from their colleagues because of cultural differences, lack of mentorship, or microaggressions that take a toll on job satisfaction and performance. Addressing these issues requires both conscious effort and structural changes within organizations.

Suggestions on continued research and data collection will help to make progress in closing the gaps. Ongoing diversity data collection also enables the piecing together of trends, as well as measuring the effectiveness of any interventions designed to drive more diverse talent through the hiring funnel. Key areas for future research include longitudinal studies tracking minority students from education through to employment, as well as the impact of specific diversity initiatives on hiring and retention rates. Additionally, organizations should be encouraged to conduct self-assessments and report data on their diversity metrics transparently.

If we put in place structured monitoring, it tells us what works and what doesn't work. For instance, studying mentoring programs specifically for minority employees can assist in determining the most effective practices all other organizations should mimic. Similarly, we can suggest that analyses of successful partnerships between higher education institutions and minority-serving businesses hold promise for examples of productive, fruitful collaboration. Policies aimed at fostering inclusive workplace environments, such as antibias training and supportive networks for minority employees, should also undergo rigorous evaluation to ensure they achieve their intended outcomes

4.4 Statistical Analysis of Minority Representation in STEM Workforce

Minority representation in STEM employment continues to be an area of interest, because it relates to social equity and innovation in fundamental ways. These dynamics also have implications for both workforce diversity and societal progress. By analyzing these statistics, we see trends and inequities among different minority groups by discipline in the STEM fields. Although there have been efforts to diversify the workplace, the extent to which minorities are represented differs greatly between demographic groups and sectors of the industry. This representation matters deeply, because it influences how innovations are created, interpreted, and propagated through society.

Through statistical analysis, this chapter explores the microcosm of employment in STEM among minorities. It looks at current trends in minority representation, highlighting sectors where there is a big disparity. The answer involves analysis of employment patterns and contrasts such as growth versus stagnation in certain demographics that illuminate ongoing challenges, the chapter adds. Sectoral details reveal where minority engagement is less or more widespread. From a broader lens, regional differences in employment statistics reveal geographic disparities impacting minorities' access to STEM careers. This chapter, therefore, seeks to contribute to the greater scholarly conversation on the barriers and opportunities in achieving minority representation in STEM disciplines by offering a comprehensive, multifaceted overview of the various factors at play and how they can inform the implementation of targeted strategies for fostering a diverse and inclusive workforce.

4.4.1 General State of Minority Representation in STEM

TABLE 4.1

STEM Minority Representation.

	% in U.S. Population	% in STEM Field	Median STEM Salary
White	60.1%	69%	$71,900
Black	13.4%	9%	$58,000
Asian or Asian American	6%	13%	$90,000
Hispanic	16%	7%	$60,760
American Indian or Alaska Native	1.2%	.3% *engineering	N/A

Note. U.S. Census Bureau, Pew Research Center, Sandia National Laboratories. From Guide to Diversity in STEM: Get an Education the World Needs, by V. McGee, 2021, ComputerScience.org.

Inequalities in STEM representation remain a topic of interest mainly due to their implication for innovation, workforce diversity, and social equity. An exhaustive review of employment data provides insight into the participation of minorities in the variety of sectors within STEM.

4.4.2 Quantitative Analysis of Minority Presence in Various STEM Sectors

The situation of minority representation in STEM, however, is quite complex and varies greatly by sector. According to the National Center for Science and Engineering Statistics, although minorities account for 31% of the U.S. population, they only make up 24% of the workforce for STEM fields (The STEM Labor Force: Scientists, Engineers and Skilled Technical Workers | NSF—National Science Foundation, n.d.). Hispanics and Blacks are particularly underrepresented in these groups compared to their share of the population in general, holding 8% and 9% of STEM jobs, respectively (U.S. Bureau of Labor Statistics, 2024). In contrast, Asians are overrepresented in STEM in relation to their share of the total U.S. population. This highlights the importance of customized strategies that meet the specific needs of diverse minority groups in STEM access and participation.

4.4.3 Growth Comparison Projection

Watching employment for the past years indicates the progress and setbacks faced by minority groups on their way into the STEM world. While there have been efforts to diversify, its growth has not been equal. The proportion of minority workers in STEM has seen incremental increases, primarily driven by policy initiatives aimed at enhancing STEM education and career access for underserved communities. However, these gains are offset by stagnant or declining trends in specific

TABLE 4.2

Employment in STEM Occupations.

Occupation category (Numbers in thousands)	Employment, 2023	Employment, 2033	Employment change, numeric, 2023–33	Employment change, percent, 2023–33	Median annual wage, dollars, 2023[1]
Total, all occupations	167,849.8	174,589.0	6,739.2	4.0	48,060
STEM occupations[2]	10,712.4	11,822.8	1,110.4	10.4	101,650
Non-STEM occupations	157,137.5	162,766.2	5,628.8	3.6	46,680

Note. Employment in STEM occupations, 2023 and Projected 2033. U.S. Bureau of Labor Statistics, U.S. Bureau of Labor Statistics, 2024, https://www.bls.gov/emp/tables/stem-employment.htm

demographics, such as Black and Hispanic workers. Examples include Black and Hispanic STEM workers with 2021 unemployment rates of 6.6% and 5.7%, respectively, compared to 4.6% and 4.5% for White and Asian counterparts (National Science Foundation, 2023). These disparities reveal persistent barriers to consistent growth in minority representation across racial and ethnic groups.

4.4.4 Identification of Industries With Highest and Lowest Minority Representation

An examination of industry-specific data uncovers differences in the minority representation of people in various sectors. Other industries, such as healthcare and technology, demonstrate relatively greater diversity. Blacks account for 11% of healthcare practitioners and technicians, a figure that is in parity with their share within the overall workforce (Funk & Parker, 2018). In contrast, in fields such as engineering and physical sciences, industries have lower levels of minority engagement. For example, women occupy only 14% of engineering roles, and many of them also belong to the minority group (Funk & Parker, 2018). Understanding these industry-specific dynamics is crucial for designing targeted interventions to enhance diversity in underrepresented fields.

4.4.5 Analysis of Employment Statistics by Region

Regional differences further complicate the landscape of minority representation in STEM. There are geographic variations in employment statistics that affect minorities' access to STEM careers. Urban centers with diverse populations, like New York City and Los Angeles, see more minorities working in STEM industries, in part because of wider access to educational resources and job opportunities. In contrast, areas that are less diverse or that lack access to a quality education have troubling gaps in representation. Non-White school districts get $23 billion less in funding than White districts despite having a similar number of students. This funding gap leads to regional disparities in educational readiness, which translates into the data in jobs given to minorities in STEM.

Statistical analysis of minority representation in STEM highlights both achievements and ongoing challenges. Quantitative measures show differences between sectors and highlight industries in which notable progress is evident and remaining work still to be done. Trend analysis highlights how incremental progress needs to be supported long term to address broader system change. Additionally, understanding regional variations in employment figures is crucial for crafting targeted policies that take into consideration the specific challenges facing minority groups in different parts of the country. However, as we delve further into these complexities, it is becoming apparent that promoting minority representation in STEM is multidimensional requiring attention toward the varied experiences and expectations of minority groups. Through this we can build a more inclusive and equitable STEM workforce, which captures the full set of talent potential in society.

4.4.6 Barriers to Entry and Advancement

At the heart of this issue is the need to not only improve individual lives but to spur on societal progress. Recognizing these obstacles is key to developing methods to create a more equitable and diverse STEM workforce. Educational gaps constitute a formidable barrier to minorities who wish to pursue STEM professions. Students hailing from communities with less-than-ideal support systems lack access to quality education, and the necessary resources to succeed in higher education and beyond are not given to them. Advanced science and math courses may be limited in schools in these areas, reducing students' preparedness for college-level STEM education (Smith, 2024). In addition, lack of sufficient funds may discourage the students from studying further at all when a university has no significant scholarship or financial aid programs. This systemic inequity in education serves as an initial barrier to entry into STEM industries, creating a cycle of underrepresentation.

Once integrated into the workforce, minorities must struggle with workplace inclusivity practices. Well, the answer is simple, a supportive workplace is crucial and many organizations still do not have this. The inclusivity practices of the organization, or lack thereof, can have a significant impact on job satisfaction and growth for minority employees. Companies with more inclusive cultures not only provide better environments for individual employees, they also achieve higher levels of innovation and productivity (Funk & Parker, 2018). In a similar vein, many workplaces today, despite having instituted diversity policies, lack effective diversity policies in their inclusivity practices.

Socioeconomic status also impacts minority participation in STEM education and careers. Socioeconomic status dictates the quality of education, extracurricular enrichment, and professional networks. Students from lower-income families face added financial strain that could push them to focus their efforts on seeking immediate work and not necessarily their ongoing education. However, the price of attending college, along with the need to manage both work and study commitments, can be exorbitant (Smith, 2024). These economic disparities make it difficult for talented candidates to be compared equally with a candidate from a more privileged background.

Systematic biases in educational institutions and workplaces also contribute to the underrepresentation of minorities in STEM. These biases show up in a number of ways, such as stereotypes that minorities are less qualified for or interested in STEM fields. These biases could deter minorities from entering these fields in the first place, fearing they would face discrimination. Further, hiring and promotion practices are often biased leading to unequal career advancement opportunities (Funk & Parker, 2018). Recognizing and dismantling these structures and biases takes intention and awareness at all levels of education and employment.

Education should be the first step in a comprehensive approach to increasing minority representation in STEM fields. In order to reduce disparities in education, ensuring access to exceptional STEM programs from an early age is essential to level the playing field. This means guaranteeing all kids access to well-resourced schools, qualified teachers, and strong curricula. Connecting students to minority professionals already thriving in STEM professions at all levels of career through mentorship programs is important to lead future generations.

BLS periodic table of science, technology, engineering and mathematics (STEM) occupations

The periodic table of chemical elements, created by Dmitry Mendeleev in 1869, is one of the most important achievements in modern science. To celebrate this achievement, BLS has created our own periodic table! Instead of elements, we have used Science, Technology, Engineering and Math (STEM) occupations. Workers in STEM occupations use science and mathematics to understand how the world works and to solve problems.

STEM GROUP

Chemistry | Geosciences
Computer Science | Life Sciences
Engineering | Mathematics
Environmental Science | Physics/Astronomy

Typical education needed for entry | Employment change, projected 2023-33 | Median annual wage, 2023 — **Ex** Example occupation

Ch — Bachelor's, 7.6%, $85k — Chemists
Cm — Bachelor's, 17.4%, $168k — Computer and information systems managers
Cp — Doctoral or professional, 2.4%, $96k — Chemistry teachers, postsecondary
Et — Associate's, 5.5%, $57k — Chemical technicians
Ce — Bachelor's, 9.8%, $112k — Chemical engineers
Me — Bachelor's, 11.0%, $100k — Mechanical engineers
So — Bachelor's, 7.2%, $68k — Soil and plant scientists
Mi — Bachelor's, 6.7%, $85k — Microbiologists
Ct — Associate's, 1.8%, $61k — Civil engineering technologists and technicians
Ep — Master's, 18.8%, $81k — Epidemiologists
En — Bachelor's, 7.3%, $95k — Environmental scientists and specialists, including health
Pt — Associate's, 7.0%, $51k — Environmental science and protection technicians, including health
Ac — Bachelor's, 21.6%, $120k — Actuaries
Ma — Master's, 3.7%, $116k — Mathematicians
Cn — Bachelor's, 11.4%, $130k — architects
Is — Bachelor's, $100k — Information security analysts
Ge — Bachelor's, 5.5%, $93k — Geoscientists, except hydrologists and geographers
Gt — Associate's, 3.8%, $52k — Geological technicians, except hydrologic technicians
Ep — Doctoral or professional, 3.9%, $106k — Environmental science teachers, postsecondary
Sa — Master's, 11.8%, $104k — Statisticians
Pp — Doctoral or professional, 3.7%, $98k — Physics teachers, postsecondary
As — Doctoral or professional, 7.4%, $128k — Astronomers
Ph — Doctoral or professional, 7.2%, $156k — Physicists
Hy — Bachelor's, 2.8%, $89k — Hydrologists

BLS U.S. BUREAU OF LABOR STATISTICS

Source: U.S. Bureau of Labor Statistics, Employment Projections program.   The BLS would like to thank the Nebraska Department of Labor for the original idea for this table.

FIGURE 4.1

BLS Periodic Table of STEM Occupations.

Note. K-12: Student resources. U.S. Bureau of Labor Statistics (n.d.), https://www.bls.gov/k12/students/careers/stem-table.htm

The efforts toward workplace inclusivity will only succeed if they facilitate an eco-system that doesn't just tolerate different opinions but leverages them. Organizations should also invest in continuous training related to unconscious bias, and, with that, adopt practices that ensure equal opportunities for growth. Supporting mentorship at the corporate level can help minority staff better navigate their way up the corporate ladder and become role models of their own.

While scholarships and financial aid can help an individual, addressing socioeco-nomic factors requires systemic change. Policies need to be established that will lead to a reduced financial burden on lower-income students. Internships and/or appren-ticeships can also be vital to ensuring that these students gain industry experience and networking opportunities that help to jumpstart their careers (Smith, 2024).

Finally, breaking down systemic biases requires persistent work throughout soci-ety. Institutions need to actively assess and adjust their policies regarding recruitment, hiring, and promotion to ensure fairness. Having clear processes, along with trans-parency and accountability from the stakeholders, will contribute to a real impact of diversity implementations. It is necessary for educational institutions, leaders in the private sector, and government agencies to collaborate in order to cultivate an equitable environment that nurtures the potential of minority talent pursuing a STEM career.

4.4.7 The Role of Diversity in Innovation and Creativity

In addition to fostering innovation, diversity in STEM strengthens workforce per-formance through improved communication and collaboration skills. This subpoint offers an investigation into how diversity can drive these results, with a thoughtful nod to the incredibly unique nature of technical fields. Studies have steadfastly shown a strong correlation between diverse teams and increased creativity in problem-solving. Diversity brings an array of perspectives that can inspire creative thinking and encourage teams to consider solutions to problems that may not be proposed in more homogenous environments (Harnessing the Power of Diversity in STEM Departments, n.d.). The fact that people on diverse teams come from different back-grounds and have different experiences and ways of thinking makes them more effec-tive at coming up with creative solutions to complex problems.

Diverse teams must engage in dynamic discussions, where conflicting ideas become valuable sources of creative tension. Such tension forces individuals to reevaluate their assumptions, leading to novel solutions. Joecks et al. (2013) noted that gender-diverse teams tend to deliver higher-quality scientific and technical out-puts. This illustrates how varying perspectives can lead to breakthrough ideas that might remain undiscovered in less diverse settings.

4.4.8 Evidence of Diverse Perspectives Leading to Technological Advancements

There's a lot of data to suggest that different perspectives contribute to techno-logical progress. Studies by the Boston Consulting Group found, for example, that organizations with diverse management teams experience 20% higher innovation revenues (Lorenzo et al., 2018). The balance of diverse cultural backgrounds among teams was significant because it came with unique cultural perspectives needed to shape new solutions.

Diversity of thought brings new ideas, questions status quo, and creates a culture where innovative solutions thrive. These differing perspectives allow for the discovery of gaps in existing technologies and create opportunities for the improvement or reinvention of products and services. Diverse cultural knowledge, for instance, can assist teams in understanding market needs the world over, helping them to create solutions that are broadly applicable, rather than falling into the trap of a narrow, one-dimensional mindset.

4.4.9 Challenges Faced by Homogeneous Groups in Adapting to Diverse Market Needs

Conversely, homogeneous groups face difficulties grasping various aspects of market needs. Without a range of perspectives, these teams may find it difficult to discern and successfully fulfil the needs of a diverse set of customers. These limitations can lead to outputs that don't connect with target audiences resulting in stunted development and lost competitiveness.

Just like our brains, homogeneous teams often fall prey to groupthink, where the desire for harmony and acceptance overwhelms disagreement and new ideas. Such conformity all too often leads to solutions that are out of sync with changing market needs. The only way to reap the benefits of diversity in organizations is to actively involve diverse talent in their processes instead of sticking with the most homogeneous teams (Zheng et al., 2013). Companies that find themselves in such positions may miss innovation opportunities, which could make them obsolete in the ever-evolving markets.

4.4.10 Guidelines for Developing Diversity Policies for STEM Organizations

With this in mind, STEM organizations can follow certain guidelines for establishing diversity policies to enable effective use of diversity. Setting specific diversity targets, encouraging inclusive recruitment practices, and providing ongoing support for underrepresented communities are some of the critical measures to take. Encouraging open communication and collaboration across diverse teams ensures that all voices are heard, fostering an environment conducive to innovation (Jones et al., 2020).

Companies can also introduce trainings that foster cultural awareness and competence among employees. This strengthens team dynamics and reduces misunderstandings, helping diverse teams work in sync toward shared goals. One such type of support is structured mentorship programs specifically designed for minorities in STEM fields, which can provide guidance and connect them with career advancement opportunities.

4.4.11 Role of Policy and Advocacy

Policies of governments and institutions are vitally important for improving minority inclusion in academic STEM. These policies are designed to create equality of opportunity by opening doors for underrepresented populations and breaking down institutional barriers. An important policy to mention is that of affirmative action programs, which has played a crucial role in promoting diversity in STEM education and employment, and equitable access to educational resources and job opportunities. As a result, we have seen fairer representation of minorities in higher education institutions.

In addition, multiple government programs are geared toward enhancing K-12 education in underserved areas. These initiatives include support for enrichment programs in science, teacher preparation and training, and infrastructure improvement. These policies directly address educational disparities early on and provide a strong foundation for minority students pursuing STEM disciplines. Inclusive curricula as well as financial aid policies are also critical for widening access to STEM education (Sustainability Directory, 2025).

Advocacy campaigns have been instrumental in driving changes in minority participation in STEM. These campaigns may include advocating for the economic and social benefits of a diverse workforce, thereby urging policymakers and organizations to prioritize diversity. For example, efforts such as those led by the American Association for the Advancement of Science and Society of Hispanic Professional Engineers have made great strides in raising awareness and garnering support for diversity initiatives. They reach out to students from underrepresented backgrounds and give them a taste of what STEM fields look like by running workshops and mentorship programs.

Nonprofit organizations play a critical role in influencing STEM education policies. They serve as advocacy groups that push for systemic changes within educational institutions and industries. Organizations like the National Girls Collaborative Project work tirelessly to empower girls in STEM by promoting gender equality and offering resources that encourage female participation. Such nonprofits collaborate with schools and universities to implement policies that create supportive environments for minority students. Their efforts are backed by research showing that diverse teams drive innovation and outperform homogeneous groups (https://www.facebook.com/plantae.org, 2024).

Lastly are scholarship programs, an essential element in fostering diversity in STEM. Scholarships are financial gifts given to students who would not have been able to pay for their higher education. Numerous offers are available specifically for certain groups of minorities—they help narrow the gap in accessibility to higher education. Scholarships for underrepresented students in STEM disciplines through programs like the one's offered by NSF may ease financial burdens and provide academic support for students to succeed.

Monitoring and evaluating progress in minority representation is essential for ensuring the effectiveness of these policies and initiatives. It involves collecting data on minority enrollment, retention, and success rates in STEM fields. These data help identify gaps and areas that need improvement, enabling policymakers to adjust strategies accordingly. Continuous evaluation also highlights successful programs and practices that can be replicated to achieve broader impact.

Another important part of increasing minority representation in STEM is involving community and education institutions. Community engagement leads to trust and collaboration between educational institutions and the communities they are meant to serve. Organizations can partner with schools and universities to create outreach programs that introduce minority students to STEM careers. Such collaborations can also lead to mentorship opportunities, providing students with role models who will guide and inspire them.

While inclusive policies and advocacy efforts are essential, their success depends heavily on societal attitudes and perceptions. Overcoming deeply ingrained biases requires a concerted effort to challenge stereotypes and promote equity. Cultural sensitivity training and awareness programs can help shift mindsets, making diversity an integral part of organizational culture. Institutions must invest in training programs that foster inclusive mentoring and leadership, ensuring that everyone feels valued and respected.

4.5 Why STEM Careers Could Offer Economic Stability for Minorities

Understanding how STEM careers offer minorities economic security unveils an area of promise and growth. With the world's industries moving toward technology and innovation, skills in STEM are ever increasingly in demand. For minority groups, this demand is an opportunity not just for single economic movement but, in reality, is also in the interest of community development. Working on these thriving fields ensures individuals can find well-paying and sustainable jobs, helping the economy flourish and paving way for the next generation. This doesn't make the road into STEM fields easy, but it does provide a combination of accessibility and adaptability with plenty of pathways to study to meet many different needs and circumstances.

4.5.1 Making High-Demand Fields More Accessible

The STEM field is rapidly expanding and offers a unique opportunity for economic stability for minorities. STEM occupations are projected to grow at a much faster rate than those in other industries. This trend is driven by advances in technology and the growing dependency on scientific and technical professionals in a variety of industries (Hawthorne et al., 2022). This expansion suggests that education and career pathways in STEM offer not just jobs but more secure careers, with the potential for personal and professional advancement.

STEM careers are some of the most exciting options out there, in part because many of them are accessible without needing to go to a traditional four-year college.

Many STEM jobs, like information technology or technical support, require skills that can be learned in associate degrees or certifications. For example, there are over 16 million more technical jobs today that require only an associates' degree or the equivalent qualification (Barone, 2019). This gives a chance to minorities who may encounter obstacles in accessing or affording full-degree programs. Community colleges play a vital role here as well, offering affordable education and often aligning curriculums with the current demands of the job market (Hawthorne et al., 2022).

Coding bootcamps and online courses, which offer work-ready skills, are gaining popularity as effective alternatives to the traditional education route. These programs aim to prepare people with targeted, in-demand skills in a short time. Most of them do offer flexible learning schedules at a much lower cost, hence they can reach a wider audience. They serve working adults or those who, for economic reasons, are unable to commit to the traditional higher education path, unlocking the pathway to a STEM career for anyone with the interest and aptitude. Coding bootcamps, for instance, can turn novices into qualified software developers in months instead of years, improving job readiness and employability.

4.5.2 The Financial Advantages of Careers in STEM

The financial advantages of pursuing a STEM career are significant, offering numerous benefits that extend beyond the immediate paycheck. At the forefront is the issue of salary. Careers in STEM often present higher entry-level salaries compared to non-STEM jobs. This is because these roles demand specialized knowledge and skills that are high in demand yet available to a relatively small portion of the workforce. As individuals acquire more experience and technical expertise, they can command even higher salaries, presenting a robust incentive for those considering these career paths (What Is a STEM Career and What Are the Benefits of Choosing One, 2025).

Another attractive feature of STEM jobs is job stability! STEM roles typically command greater job security than most other jobs, thanks to the technical expertise they require. As the world moves toward an economy rooted in technology and innovation, these skills become indispensable, and STEM professionals find themselves less susceptible to downturns in economies and fluctuations in job markets. But during such times (or any time, really) when industries work to streamline operations or pivot strategies to work through challenges, the need for capable STEM workers still persists since their responsibilities are typically elemental to operations and innovation.

Data from U.S. Bureau of Labor Statistics, for example, underscore that STEM occupations grow faster than those in other sectors and are more resistant to employment volatility. With the rapid pace of change today, where entire industries can change overnight due to technological advancements or consumer demand shifts, these benefits are vital.

STEM provides numerous opportunities for career advancement. Many STEM careers have a clear structure in which one can climb the ladder of success simply by improving one's skills, working on projects, and furthering one's education. As they go along with their careers, professionals advance toward higher-level roles—be it senior roles, management roles, or areas that require more in-depth knowledge. These strides often lead to significant jumps in their salaries, which have helped cement STEM careers as financially lucrative options (The Benefits of a Career in STEM Education—STEAMspiration, 2023).

In addition, jobs in the STEM fields tend to offer the most benefits, adding to stability over time. These packages often include health insurance, retirement plans, stock options, and bonuses, all of which add considerable value to the total compensation. This is reflected in the fact that technology companies with the most competitive salary structures also offer lifestyle benefits to ensure their employees and their families maintain a good quality of life and financial well-being.

Comprehensive benefits are particularly attractive because they provide a safety net that extends beyond the individual employee, offering peace of mind through coverage of unforeseen health issues and the assurance of a secure retirement. These benefits reflect the value organizations place on retaining talented professionals who contribute to their growth and innovation.

Economic stability achieved through STEM careers positively reverberates throughout local economies. Individuals who secure jobs in technology, engineering, or scientific research typically enjoy higher salaries and job security compared to those with non-STEM occupations (Okrent & Burke, 2021). This economic uplift not only benefits individual households but also contributes to the overall financial well-being of their communities. With greater disposable income, families invest more in local businesses and services, thereby enhancing the economic vibrancy of their neighborhoods. Additionally, increased employment opportunities derived from STEM engagement can lead to reduced poverty rates and improved standards of living, thereby creating a virtuous cycle of prosperity and growth within minority communities. The American dream is tied to an education and a good job, and with high-demand careers on the rise, engaging young students and exposing them to the world of STEM is of utmost importance.

4.6 Entrepreneurial Opportunities in STEM for Marginalized Groups

An exciting avenue for these benefits lies in the realm of entrepreneurship and innovation within STEM fields. By leveraging their technical skills and knowledge, individuals from marginalized backgrounds can establish ventures that not only secure their economic independence but also drive broader social change.

A standout trend includes the emergence of technology startups with founders from previously underrepresented groups. These entrepreneurs are pushing through boundaries, coming up with great companies that people never thought possible. Another type of opening is that of minority-owned businesses in areas like software development, cyber security, and biotechnology. Part of the rise is also because marginalized people come with different perspectives and solutions to problems, which can make for unique and valuable innovation. In addition, the explosion of the digital economy has drastically lowered barriers to entry, enabling everyone from anywhere in the world to start and then scale tech ventures.

Access to funding is a critical component for the success of these startups. Various avenues exist to support minority-owned businesses, including grants, venture capital, and crowdfunding. Grants like those offered by Women Who Tech provide essential financial resources to women-led startups, addressing a significant gap given that only 2.8% of venture capital goes to women-led ventures (*Home | Women Who Tech*, n.d.). These grants, along with similar programs, play a crucial role in leveling the playing field and enabling diverse entrepreneurs to compete effectively. Moreover, venture capital firms focusing on minority entrepreneurs, such as RareBreed VC and Passbook Ventures, are becoming increasingly prevalent, offering both funding and strategic guidance.

Similarly, people can use crowdfunding platforms to raise funds much easily. These platforms offer entrepreneurs the chance to test their business concepts, while at the same time attracting investment from a diverse pool of backers. Crowdfunding campaigns for successful minority-owned ventures like innovative technology solutions or community-minded projects can hold particular appeal with funders who back founders from diverse backgrounds.

This is also where mentorship programs play a critical role in cultivating tomorrow's minority STEM entrepreneurs. The Inclusive Teaching Academy initiative at Eastern Washington University and the Native Explorers Program initiative at Oklahoma State University are examples of mentoring approaches designed within academic contexts to create opportunities for growth (Staff, 2022). They offer mentorship, help participants to grow, and provide a powerful networking platform to connect mentees with the people spearheading their industry or even their future investors. Relationships are everything in the maze of entrepreneurship and scaling a business.

Mentorship may also involve sustained support for the business as entrepreneurs adjust their business models and work on leadership and challenges that are specific to these environments. Having mentors to guide one on market trends, strategic decisions, and emotional sustenance is a must for sustaining the success in the long term. Specific programs tailored for minority entrepreneurs are critical to overcoming systemic obstacles and providing culturally competent advice to boost their impact and effectiveness.

Minority STEM entrepreneurs still face significant barriers to success. Challenges around how to enter the network, gain initial funding, and secure their fair share of resources remain obstacles. But the good news is that these are surmountable barriers, with concrete solutions. One effective approach is the creation of incubators and accelerators tailored to minority-owned startups. These programs offer comprehensive support, including office space, technical assistance, and investor connections. For instance, some nonprofits and community organizations run incubators that specifically support underrepresented founders, providing them with the tools they need to succeed.

One also needs to be an advocate for changing policies that support greater D&I within the startup ecosystem. To open doors to minority-owned businesses and make them a reality, it is essential that a plan be put in place for policies such as investment, tax incentive, or separate funding pool. Certain partnerships between the private and public sector can help create an atmosphere where minority entrepreneurs flourish.

The level of education and further training is also essential. Business skills, financial management, and legal issues workshops, webinars, and skill-building teach entrepreneurs what they need to know about the seamier side of starting a business. Collaboration with industry experts can also provide educational institutions with content that is both practical and inspirational.

Another solution involves creating a supportive community where minority entrepreneurs can share experiences, resources, and advice. This could be through networking events, online forums, or peer mentoring schemes where entrepreneurs come together to form relationships and work on projects while gaining support on the challenges they face. Strong communities of practice can support others by providing these values, critical in maintaining the right level of motivation and resilience needed for a successful entrepreneurial journey.

4.7 Programs Aiding Minority-Owned and STEM-Based Businesses

Without federal and state programs that support minority-owned STEM businesses we cannot expect for them to flourish as they should. These programs are developed to give financial support, tools, and opportunities that enable those business in advancing. For example, the NSF offers grants and supplemental funding specifically aimed at broadening participation in STEM fields. Programs like the Historically Black Colleges and Universities—Undergraduate Program (HBCU-UP) and Excellence in Research (HBCU-EiR) are targeted at strengthening STEM education and research capacities at HBCUs. In particular, HBCU-UP awards funds to enhance undergraduate STEM education, while HBCU-EiR provides support for developing research capacities (Supporting Black/African Americans in STEM—Broadening Participation in STEM | NSF—National Science Foundation, n.d.).

State-level support in addition to federal initiatives play a major role. In fact, several states have even created offices or agencies specifically tasked with promoting minority-owned businesses with grants, tax incentives, and business development assistance. State efforts are often the most crucial, bringing support to communities with particular needs.

Nonprofits are also impactful in sparking STEM entrepreneurship among disadvantaged minorities. They offer resources and mentorship, as well as networking opportunities. They host regular workshops, seminars, and conferences for minority entrepreneurs in STEM to expand their skills and knowledge. Many nonprofits, in fact, are essential intermediaries linking minority business owners to capital resources, other firms, and institutions of higher learning.

Nonprofit workshops and mentorship are great for personal growth and development. These organizations tend to create communities for minority business owners to come together, exchange stories and learning from one another and industry experts. With this, they will be able to gain a considerable understanding about how things are done thereby leading them in to doing what's best for business development and management and increasing sales chances.

Also, many educational institutions have incubator programs that cater specifically to minority-owned businesses. Incubators offer minor resources, including office space, funding, and technical support for entrepreneurs to begin their ideas and businesses. Additional support from educational institutions could yield specialized training and certification programs designed to train minority business owners with the skills needed in the ever-changing STEM industry.

Resource networks and coalitions are powerful tools to help minority STEM businesses. Organizations such as the Minority Business Development Agency (MBDA) offer programs in business consulting, procurement assistance, and financial services. These networks support minority entrepreneurs in overcoming the challenges associated with creating and building a STEM business.

At the same time, coalitions like the National Minority Supplier Development Council (NMSDC) link minority-owned businesses to supply-chain diversity-seeking organizations. Joining such networks can offer access to high-value contracts and partnerships, which has the potential for exponential business growth.

Additionally, online platforms and forums build on these resource networks because they offer virtual places where minority entrepreneurs can meet each other in order to exchange information and ask for advice. These kinds of digital communities are especially helpful for active users who do not have a support network where they live. These sessions often include dialogue around practices, trends, and collaborative opportunities within a specific industry.

While highlighting existing programs and resources is essential, it's equally important to provide guidelines for accessing these support systems, like the following:

1. **Identify Relevant Programs:** Understand which federal, state, and nonprofit programs align with your business goals. Research eligibility criteria and application processes thoroughly.

2. **Leverage Educational Partnerships:** Engage with local universities and colleges to explore potential partnerships. Participate in internship programs and collaborative research projects to gain access to valuable resources and expertise.

3. **Utilize Resource Networks:** Join networks and coalitions that offer support to minority-owned STEM businesses. Take advantage of consulting services, procurement assistance, and financial planning support.

4. **Participate Actively:** Attend workshops, seminars, and networking events organized by nonprofits and resource networks. These events provide opportunities to learn, network, and gain visibility in the industry.

4.8 Final Insights

This chapter has focused on the financial security, career trajectory, and economic empowerment that can come from having a STEM job, especially for those who have been historically underserved. We have looked at comparing STEM fields with other industries to see how higher salaries and job resilience add to long-term economic security.

Indeed, this conversation is a poignant reminder of the need for intentional programs and systemic support for growing diversity in STEM. Mentorships, scholarships, and inclusive educational practices are all important programs that aid in addressing equity gaps. They contribute to a healthier workforce, drive innovation, and help build community resilience. This chapter reinforces the importance of ongoing work to support greater opportunity in STEM for the underrepresented groups so that all can benefit from the value of a career in this area.

5

Enriching the Culture Boosts Bottom Lines

5.1 How Diversity and Inclusion Programs Can Affect Company Culture and Business Outcomes

The benefits of D&I efforts are truly game changers in culture and business trends. These programs enhance the workplace culture to be more inclusive and fun, which results in smarter collaboration and creativity among employees. A workplace that values and respects diverse perspectives is better positioned to solve more complex tasks, giving the company a stronger, more integrated team.

In the following chapter, we explore how D&I initiatives drive better business metrics like employee retention, job satisfaction, and innovation. The series of articles will cover the relationship between diverse teams and their capacity to fuel creativity, problem-solving, delivering better products/services, etc. The chapter will explore from both a human capital and operational perspective, using case studies and data-driven analysis, what the benefits are of building an inclusive culture in organizations. It will also introduce some practical tips to aid the successful execution of diversity initiatives, discussing the necessity for reassessment and adaptation over time.

5.2 The Effects of STEM-Based Educational and Workforce Initiatives

Because the future progress of life in this or any country relies on a fair shot at education, national STEM initiatives that emphasize inclusivity are equally crucial. These initiatives seek to provide access to STEM education to demographics that have been left out of the industry, such as Latinx, Indigenous, and Black/African American students (Palid et al., 2023). These initiatives serve to broaden the talent pool by being more inclusive and ensuring that people from nontraditional paths have a chance at engaging in STEM employment. It is especially significant in light of the fact that college students from these backgrounds receive a smaller share of bachelor's degrees in STEM disciplines versus their general population demographics (Fry et al., 2021; NCSE, 2021).

DOI: 10.1201/9781003453581-5

Government and private sector investments play a crucial role in advancing these diverse STEM programs. When both sectors invest in inclusive STEM initiatives, the impact extends beyond education into technological advancements and innovations. For example, programs funded through grants and mentorship opportunities such as those from the NSF support underrepresented groups in STEM (Advancement of STEM Graduate Education: Diversity, Equity, Inclusion and Accessibility, 2024). These investments not only promote equity through diversity but also spur technological advancements by creating an environment that values diversity. This is a key part of the value of diversity in teams—it creates many different ways to approach problem-solving, providing more creative and effective solutions.

Collaboration of higher education with industry plays an equally important role in making students job-ready by equipping them with hands-on experience. The alliance provides the students with learning-by-doing experience, which helps them in connecting theoretical knowledge to real-time application. Most schools partner with industry to offer internships, apprenticeships, and co-ops to give students a work-based view of their future careers. These partnerships deliver a comprehensive education where the knowledge and skills students acquire translate into valuable insight when they enter the workforce.

Recommended pathways for developing these partnerships can be to define shared objectives, align the content of academic curricula with industrial requirements, and allow for feedback systems that help to measure and refine these partnerships. Additionally, involving industry professionals in the educational process through guest lectures, workshops, and mentoring can further enhance the learning experience for students.

Inclusive STEM programs have also been shown in longitudinal studies to increase the retention and job life satisfaction of minority students (Estrada, M., Hernandez, P. R., & Schultz, P. W., 2018). Longitudinal studies of this type, which follow students for relatively long periods and as such offer detailed data on the impact of interventions, are somewhat rare. For example, studies show that underrepresented students in STEM who participate in inclusive programs are more likely to graduate and enter successful careers in their chosen fields. Programs that are intentionally supportive of minority students usually provide opportunities for mentorship and financial assistance, as well as community-building events that lead to improved retention rates and overall satisfaction.

A case in point is the generation of STEM Intervention Programs (SIPs) across postsecondary institutions. These programs target the recruitment, retention, and success of underserved students in STEM (Rincón & George-Jackson, 2016). Through SIPs, participants are specifically trained to navigate the challenges of competitive academic environments and financial obstacles faced by minoritized populations. Such programs offer individualized aid that can help students from many different backgrounds who need encouragement and assistance in order to flourish.

In addition, long-term support of government policies and structures supporting these groups are necessary. This moves toward systemically embedding D&I in STEM education and workforce development through targeted policy changes. For example, the NSF partners with federal agencies, colleges and universities, and policy organizations to influence and construct policies that encompass diversity, equity, inclusion, and accessibility among participants in STEM fields (Advancement of STEM Graduate Education: Diversity, Equity, Inclusion and Accessibility, 2024).

Engagement and outreach inclusivity work also help to establish inviting environments in STEM. STEM-career inspiration programs for underrepresented populations frequently have partnerships that extend to museums, science centers, public schools, and the media. Such efforts allow learners to encounter STEM beyond standard classroom environments, thereby increasing the reach and appeal of the field.

Additionally, mental health career grant opportunities in STEM are needed to support graduate professional development. Programs like the Mental Health Outcomes and Professional Experiences (M-HOPES) project at Montana State University Billings recognize the specific challenges that come with being a graduate student in STEM, providing mental health resources and advocacy to meet their distinct needs (Advancement of STEM Graduate Education: Diversity, Equity, Inclusion & Accessibility 2024). Through such initiatives, emotional intelligence is evidently being considered as an add-on feature that enables overall success and retention in STEM projects toward academic or professional development.

Scalable storytelling interventions, such as those implemented by Boise State University, leverage the power of sharing personal narratives to support graduate student success in STEM. These interventions involve sharing experiences and strategies for overcoming obstacles, fostering a sense of community and resilience among students (Advancement of STEM Graduate Education: Diversity, Equity, Inclusion and Accessibility, 2024). By promoting an inclusive and supportive culture, storytelling interventions help destigmatize failure and encourage persistence in the face of challenges.

Programs such as the one at University of Washington look to improve teaching practices with evidence-based practices, understanding that effective teaching is crucial in higher education (Advancement of STEM Graduate Education: Diversity, Equity, Inclusion and Accessibility 2024). In short, these initiatives help create a more supportive and effective learning environment for all students by helping to have educators ready to implement inclusive, engaging teaching methods.

5.3 A Closer Look at Minority Participation in STEM

Before we can understand the impact of lack of minority presence in STEM and what it means for organizational success, we have to look at the current state of things. Those demographics form the baseline from where, and only from where, an organization can then test for targeted diversity. Analysis of these data acts as a guide in effective targeted inclusive hiring practices for companies. For example, if an organization notices a significant underrepresentation of Hispanic professionals in engineering roles, it might initiate outreach programs at universities with high Hispanic enrollment to attract talent from these groups.

Barriers to entry into STEM fields are multifaceted and have a profound impact on retention and satisfaction among minority professionals. Some of these barriers include the lack of access to quality education, limited exposure to STEM careers during early education, and financial constraints. Additionally, workplace culture plays a critical role. Minority professionals often face microaggressions, implicit biases, and even overt discrimination that can make them feel unwelcome and under-valued. This hostile environment can lead to higher turnover rates among minorities. Addressing these cultural issues requires systemic change, including training for all employees on D&I and creating support networks for minority employees.

Perceptions within societal evenness additionally play essential roles in minority enlistment. On the hiring side, these biases bleed into real issues on both how prospects perceive applicants and the experiences of professionals once they have been brought in. For example, there's a prevalent stereotype that certain races are less good at math and science work directly influencing choices of recruiters or behaviors of the surrounding colleagues. It is essential to have educational programs that work against these stereotypes. Companies can also collaborate with schools to deliver awareness and outreach programs that showcase successful minority STEM gradu-ates. This approach not only educates the broader society but also provides role mod-els for young aspirants from underrepresented backgrounds.

To increase minority representation in STEM, strategies must be multifaceted and sustained. Mentorship programs have demonstrated promise in being able to help professionals of color. A new hire is paired with experienced mentors who use their wisdom and knowledge to provide support, guidance, and networking opportunities that are crucial to progressive career moves. Studies have indicated that mentees, particularly from minority backgrounds, benefit significantly from such relation-ships, feeling more integrated and supported within their organizations (National Academies of Sciences, Engineering, and Medicine et al., 2019). Moreover, peer mentorship adds another layer of support, helping to normalize the struggles faced by minority professionals and providing a sense of camaraderie and belonging.

Another measure is scholarships aimed at minority students. Having financial aid helps eliminate a major obstacle to STEM fields. And they can help fund scholar-ships that are connected to internships and co-op positions within their companies, offering financial support and hands-on work experience. By adopting this dual-pronged process, the program attracts more bright minds and equips these students internally upon their first day of work, so they will already be trained for whatever comes next—meaning they are less likely to leave after graduation.

Diversity measures should be continuously assessed. Employee satisfaction sur-veys, retention rates, and the breakdown of who holds the various jobs in terms of demographics can be metrics that help a firm see how its diversity efforts are working. The feedback generated should be used to inform a continuous process of improvement and adaption of strategy, so organizations can ensure their initiatives maintain relevance and receive returns over time.

5.4 Breaking Down the Psychological Barriers for Minorities in STEM

Eliminating psychological barriers for minorities in STEM is essential not just for individual success but also for fostering a more inclusive and innovative scientific community. These barriers often manifest as stereotypes, biases, or lack of representation, which can hinder minorities from fully engaging or excelling in STEM fields. By addressing these issues, the path is paved for a more diverse range of thoughts and experiences that enrich scientific inquiry and technological advancement.

When it comes to inspiring future generations, representation matters; hence, the presence of diverse racial, ethnic, and gender identities in STEM is critical. Children and young adults are more likely to follow career paths in fields where they see themselves succeed. Many have said that increased representation means that there are powerful role models on which the aspirations and educational paths of those students, who are likely minorities themselves, can depend. When one sees someone from one's background who has made it in STEM, not only does it motivate one, it also gives one the confidence that one can achieve this, too.

Also central to this change is enabling people to have faith in themselves. Getting them past the barrier of stereotypes and bias allows them to see themselves as capable and deserving contributors in STEM fields. It encourages them to take ownership and responsibility of their work, that is the key to grow professionally. When individuals perceive themselves as competent, they engage more deeply with their work, demonstrating higher levels of determination and resilience in the face of obstacles. The transactional theory of stress and coping suggests that an individual's appraisal of their ability to cope is just as important as the stressor itself (Assari, 2016).

Improved self-efficacy, a by-product of the removal of psychological barriers, is a key component of superior performance and persistence. Self-efficacy is the belief in one's ability to successfully perform tasks. Those with greater self-efficacy are likely to be more engaged and to persevere more when they encounter challenges. They see themselves as competent. Hence, by creating favorable conditions for developing self-efficacy, we empower minorities to confront challenges more effectively.

This new sense of confidence is also essential for risk-taking, and for embracing innovative ideas. In STEM specifically, creativity often arises when people feel free to pursue potentially unorthodox approaches to the problems at hand, and to suggest new products or features without fear of criticism or failure. A supportive atmosphere encourages minorities to take calculated risks, contributing to groundbreaking discoveries and advancements. Confidence is contagious, and as minority individuals demonstrate success in these areas, it inspires others to follow suit, gradually transforming the cultural landscape of STEM disciplines.

Removing psychological barriers supports personal growth and professional advancement. As individuals grow more assured of their skills, they become more proactive in seeking opportunities for advancement. Whether it's taking on leadership roles, pursuing further education, or engaging in complex research projects, the possibilities expand significantly when psychological impediments are eliminated. The enhanced belief in personal capability can be particularly empowering for minority groups traditionally underrepresented in leadership positions, breaking ceilings and setting new precedents.

Retention rates also improve notably when psychological barriers are removed. Motivated workers who feel appreciated and recognized for their unique talents are likely to stay committed to their positions. In creating a culture that appreciates the contributions of all team members, companies tend to keep their talented workers, saving the companies the costs involved in turnover (such as training or facilitating new hires). Workers feel inspired to give their very best, and career-oriented workers want to be in it for the long term (Whittaker & Montgomery, 2012).

The ability to conquer psychological obstacles leads to a strong and flexible workforce poised to face the next challenge. Changes in the scientific and technological landscape are happening very fast, so one should be able to adapt. When team members have faced and overcome psychological barriers, they tend to be better prepared to deal with new developments and unexpected changes. This resilience becomes a catalyst for growth and adaptation because it allows organizations to quickly pivot to new methodologies or technologies.

Additionally, a psychologically supportive atmosphere lowers the stress and anxiety levels of minority group members and frees them up to focus on their personal and professional development. The supportive culture fosters a commitment to continuous learning and innovative thinking, empowering them to make even greater impacts in STEM fields. These environments also develop role models within communities, who motivate the next generations to pursue careers in STEM and diversify the talent entering such workforces.

5.5 Entry Opportunities Into STEM for Underrepresented Communities

Entry-level STEM programs specifically targeting underrepresented groups are essential in creating awareness and interest in STEM careers. These programs often begin at the high school level, like Columbia University's Engineering the Next Generation (ENG) initiative established in 2017. ENG targets rising high school seniors from underrepresented ethnic and racial backgrounds, offering them practical research experience with engineering researchers while developing their academic and professional skills. By engaging students early, such programs can cultivate a lasting interest in STEM fields and provide a foundation for future educational and career pursuits.

Internships and apprenticeship programs are also major pipelines for these opportunities in STEM fields, bridging the skills gap and ensuring candidates have long-term sustainability. An example of internships is that offered by the Columbia University and Amazon Summer Undergraduate Research Experience (SURE) Program. They offer underrepresented participants valuable research experience and mentorship. The 10-week experience also includes free room and board, travel reimbursement, and a stipend, as well as creating avenues into a pipeline for graduate engineering programs. The STEM Guide Program at DePauw University similarly aids in the mentorship of ethnic and racial minorities who historically have been excluded by offering opportunities to these students as teaching assistants in introductory courses. They are designed to give students hands-on experience, build vital skills, and links to professional networks that will raise their employability and career prospects.

Community outreach and engagement play a significant role in demystifying STEM careers and encouraging youth interest through collaborations with local schools. The Girls STEM Institute (GSI) at Indiana University–Purdue University Indianapolis focuses on providing holistic learning opportunities for girls and young women of color, aged 9 to 18. Through culturally grounded, inquiry-based, hands-on STEM curricula, GSI aims to instill interest and confidence in pursuing math and other STEM careers. Additionally, community outreach initiatives like the STEM Scholars Program at Iowa State University expose students to diverse STEM disciplines and help them understand how these fields connect to social issues. By engaging with local communities, these programs can inspire future generations to pursue STEM careers and contribute to the diversification of the workforce.

Technological advancements in recruiting have streamlined the process and expanded access to STEM opportunities and been supported by digital literacy programs. Stanford University's Inclusive Mentoring in Data Science Program matches undergraduates from underrepresented backgrounds at a large research university with graduate student mentors for free, online, one-on-one data science mentoring. Participants receive coaching in planning their course of studies, navigating internship opportunities, and preparing applications, as well as tutoring in specific aspects of data science. Programs like this leverage technology to reach a broader audience and provide tailored support to underrepresented students, helping them overcome barriers to entry in STEM fields.

Similarly, Foothill College's Science Learning Institute (SLI) has introduced a number of creative initiatives to promote equity in STEM opportunities for all students and persons from underrepresented communities aspiring toward educational success and STEM careers. For instance, programs such as the Program for Readiness and Exploration in STEM (Pre-STEM) introduce students to careers in STEM disciplines, connect them with social justice issues, and bridge the gap between two-year colleges and four-year institutions. The Pre-STEM Summer Institute provides a free three-week academic summer experience for new students to develop exposure to data science, calculus, and college-level topics. This is an example of how technology and digital knowledge can be used to engage the workforce and introduce individuals into the STEM careers in an inclusive way.

5.6 Findings From Corporate Measures of Diversity and Inclusion Efforts

When examining the impact of D&I initiatives on company culture and business outcomes, it's crucial to start by looking at data from surveys and studies that reveal corporate practices and their effects. These data-driven insights illuminate how inclusive efforts shape various aspects of an organization, influencing everything from employee morale to innovation metrics.

Survey results show that strong diversity programs lead to higher employee morale and retention rates across the board as well, but they have the biggest impact among underrepresented groups. These firms have a 35% higher probability of yielding above-average financial returns per McKinsey's 2020 report (psico-smart.com, 2020). This correlation indicates that employees feel more appreciated and a part of the team, making them more satisfied with their roles and less likely to leave when they see a genuine commitment from management toward diversity. When we create energetic work cultures, our people become excited to come to work. That results in lower turnover and better performance.

The way employees perceive D&I measures is also key to understanding how effective D&I has been. These perceptions can then be captured via surveys that give invaluable insights into what is going well and what might need tweaking. For example, in a Glassdoor study, 67% of job seekers believe diversity is important when evaluating companies (humansmart.com.mx, 2024). This cultural inclusivity can only be achieved through listening to those voices of the employees. This regular dialogue not only aids in fine-tuning current programs but also in establishing trust and openness, which boosts worker morale exponentially.

The connection between inclusivity and innovation metrics provides another compelling argument for investing in diverse teams. A Boston Consulting Group survey found companies with more diverse management teams have 19% higher revenues because of innovation (psico-smart.com, 2020). Regardless of industry, diversity among teams adds and impacts new ways of thinking/problem-solving by bringing different points of view to the table. For example, a Fortune 500 company made changes so that teams were more diverse culturally, which resulted in 31% higher job satisfaction and 20% better productivity. These data also illustrate that inclusivity creates the supportive environment for innovation, leading to benefits of a real business competitive edge and financial performance.

Comparing companies with inclusivity programs to those without them also underscores the business case for diversity and highlights the risks of exclusion. Firms with comprehensive D&I initiatives often outperform their less diverse counterparts. McKinsey's research shows that organizations in the top 25th percentile for gender diversity on executive teams are 21% more likely to experience above-average profitability (humansmart.com.mx, 2024). On the other hand, companies lacking such programs may face increased challenges related to recruitment, employee engagement, and overall performance.

Take two tech companies, for instance, one that values diversity and offers mentorship programs and regular inclusivity training, and one that does not. After one year, the first company saw a 15% rise in employee satisfaction and experienced a 20% increase in product development efficiency. The other company in contrast, had high turnover rates and struggled to engage staff. The hugely differing results show how diversity policies don't just help to improve a company's internal culture but also lead to real business benefits.

Additionally, the sheer connection of diverse teams with creativity alone makes it even more essential to create an environment that is suitable for inclusiveness. A Deloitte study validates this point as well, reporting that inclusive teams perform 80% better in team-based assessments than their peers, highlighting how powerful divergent perspectives can be on the performance of a team. When an employee senses respect and appreciation, they will put more effort into their work with creativity and resourcefulness. This drives both personal and company-wide success.

Inclusive practices stretch far beyond the immediate benefits of innovation or employee satisfaction and have large-scale implications on long-term sustainability and growth. With the global workforce becoming more diverse, companies that prioritize diversity will be best positioned to attract and keep top talent. Being inclusive is not only the right thing to do, it is also a business edge that one must have in order to survive in the current market.

A survey in 2017 showed that companies with the most inclusive practices have higher profitability by 21% while benefiting from high employee engagement (according to a Gallup report). That data confirms the argument that D&I efforts are not about ticking boxes or satisfying external expectations to do the right thing. They reveal themselves as critical components in a collective journey of building teams and companies fit for purpose. Envision a workplace that celebrates every voice. Not only does this culture improve morale, it also moves the organization forward to reaching its goals for success in this area.

In conclusion, the examination of data from surveys and studies reveals a clear and compelling case for the integration of D&I initiatives within organizations. Higher employee morale and retention rates, coupled with enhanced innovation and profitability, demonstrate the tangible benefits of these efforts. Employee perceptions provide valuable insights into the efficacy of current strategies, helping organizations refine and improve their approaches. By comparing companies with and without inclusivity programs, it becomes evident that the risks of exclusion far outweigh any perceived costs of implementing these initiatives.

5.7 Inspirational Examples of Diverse Teams Enhancing the Bottom Line

Market success and improved performance can be directly attributed to diversity-focused hiring practices, as many case studies show. One notable example is the global technology firm SAP. By prioritizing diversity in its hiring strategy, SAP has seen significant market growth. SAP launched an initiative to hire individuals on the autism spectrum, leveraging their unique skills in data analysis and problem-solving. As a result, the company not only created a more inclusive workplace but also enhanced its innovation capabilities, resulting in better products and higher customer satisfaction.

Unilever is another good example of this, as the company placed an emphasis on gender diversity throughout its global organization. Unilever created programs, such as flexible work hours and extended parental leave, that allowed them to maintain many of their top female leaders. By following this strategy, five years later they increased the women in leadership roles by 31%. This positive spiral effect led to a more balanced decision-making apparatus and wider viewpoints across the organization, which ultimately drove better financial performance.

Metrics also show that productivity increases with diversity on a team. Companies in the top 25th percentile for gender diversity are 25% more likely to have above-average profitability than less diverse companies (McKinsey & Company, 2023). Likewise, organizations with racially and ethnically diverse teams are 35% more likely to outperform industry norms. These statistics drive the point home that collaborating with multifaceted teams leads to an array of different thought processes, hence, innovation and resultant higher productivity and revenue growth.

Initiatives are associated with higher revenue growth and greater diversity in leadership. Research from the Boston Consulting Group suggests that innovation revenues are 19% higher in companies with more diverse management teams. That is especially important in many industries as new ideas and approaches are significant to keep a competitive edge. As leadership teams grow more diverse in terms of the experiences and cultural backgrounds they reflect, organizations are better able to tackle problems from different perspectives resulting in smarter solutions.

Testimonials from leaders who have embraced diversity initiatives provide valuable insights into the importance of diversity from both personal and professional perspectives. Consider, for example, the remarks of Microsoft's CEO, Satya Nadella, who has spoken out frequently in favor of diversity. He feels that being focused on inclusion and bringing in diverse perspectives is not just a business driver but also relates to happiness at the work place. Happy customers will fuel fast growth and help keep them. One needs satisfied customers first. In interviews and public statements, Nadella has stressed that diverse teams are more empathetic and better equipped to understand and serve a global customer base, ultimately contributing to Microsoft's sustained growth.

Indra Nooyi, the outgoing CEO of PepsiCo, has noted the revolutionary progress she made during her period in that position as a result of diversity. She also prioritized creating an inclusive culture and building a diverse leadership team. She added that this technique allowed for inventive product developments and wider market penetration. Those efforts lead to Pepsi acquiring 78% more market share in emerging markets, which clearly demonstrates how diversity is a business imperative for growth and success.

Acknowledging challenges and lessons learned is crucial for a comprehensive understanding of the benefits of diversity. Many organizations face obstacles such as unconscious bias and resistance to change when implementing diversity initiatives. However, addressing these challenges head-on can lead to meaningful progress and sustainable improvements.

IBM had its share of stumbles in the journey to workplace gender diversity, with several previous missteps. The company first grappled with deep-seated biases and a lack of representation in leadership roles. To counter this, IBM brought in broad-based training to raise awareness on unconscious bias and build inclusive behaviors. They even created mentor schemes to help women get ahead in their careers. This contributed to female representation in leadership roles scaling, and the company building a reputation as a diversity role model.

All of this discussion embodies the continued need for reflecting on lessons learned to fine-tune diversity strategies and implementation of diversity initiatives through a continuous improvement lens. For example, Google decided it was not enough just to hire more diverse staff and had to do the work of actually making sure everyone is included. This understanding led the company to concentrate on creating a more open communication culture that gave minority employee groups space and permission to share their true feelings. This all-encompassing method has helped raise employee happiness and motivation levels, which helps in securing Google's position as a leader on diversity.

5.8 Bringing It All Together

This chapter provides real-world examples from recent research of how D&I initiatives can have some effect on company culture and business success in general. An emphasis on the benefits of diversity groups with a wider range of perspectives could unite over solutions, inspiring creativity and innovative ideas. In this chapter, we further discussed the preceding concepts by providing direct financial information that illustrates how investments made on behalf of both government and private sectors to support a more diverse STEM curricula would inevitably serve well beyond just an educational perspective and stretch across the playfield from technology developments to retention rates for minority students.

In addition, by providing actual case study and statistical data to substantiate the business impact of introducing inclusive approaches, for example, about companies such as SAP and Unilever, this chapter showed how market growth and innovation capabilities are increased with a focus on diversity. This analysis shows that diverse leadership leads to higher revenue growth, reinforcing the idea that inclusion is not just the right thing to do, but a significant business advantage. By adopting these initiatives, companies can foster the kind of environment that draws top talent, increases employee satisfaction, and leads to long-term success.

6

The Rapid Evolution of STEM

How to Account for Everyone

6.1 Approaches to Growing and Sustaining a Diverse STEM Workforce

Creating and nurturing an inclusive, sustainable STEM talent pipeline is a many-layered process. Having a talent pool in STEM that is diverse is critical in promoting innovation and enables us to solve increasingly complex problems from more viewpoints. We need to realize that it is important for organizations to attract, develop, and retain people from all over the world as a driver of creativity and problem-solving. The unique focus of this chapter is on the tools and methodologies that should be employed in order to develop and maintain such an inclusive STEM workforce, so as to ensure that individuals with talent across all populations may help to shape these crucial fields.

This chapter helps readers work on creating an inclusive ecology in STEM institutions. It explores historical context, bringing to light the remarkable achievements of many scientists and engineers responsible for creating and developing these areas in the face of long odds. The chapter also showcases current approaches to recruitment in STEM, such as outreach activities, mentorship programs, and addressing financial barriers. It also covers the practical ways in which effective DEI programs can support underrepresented employees. The chapter profiles successful community-based initiatives and organizational frameworks as a guiding beacon for those hoping to enact lasting change within their own institutions.

6.2 Existing Contributions to STEM From a Historical Perspective

Many different people have been essential in constructing STEM fields over the years. Not only does highlighting these contributions put an exclamation point on why representation is so crucial, but their stories should serve as inspiration for generations to come. One way to counteract the perception of science and technology as

DOI: 10.1201/9781003453581-6

an exclusive arena is by studying the victories of accomplished figures from a range of diverse backgrounds.

An African American agricultural scientist who was born into slavery, George Washington Carver, is a noteworthy example. He went on to become Iowa State Agricultural College's first Black student and eventually changed farming by exploring over 300 uses for the peanut. His research provided the basis for ecologically based farming systems, and demonstrated the untapped capabilities of marginalized groups in scientific innovation.

Marie M Daly, the first Black woman to receive a Ph.D. in chemistry in the U.S., contributed greatly to the field of biochemistry. Her research had influence on reducing heart disease by targeting cholesterol. The pioneering work of Daly provides a model for how breaking barriers in STEM can contribute to broader scientific and technical breakthroughs that advance all of society.

Katherine Johnson, renowned for her work as a "human computer" at NASA, played a pivotal role in the early space missions, including calculating the trajectory for the first U.S. manned spacecraft and the Apollo moon landing. Johnson's work challenged gender and racial norms in STEM, demonstrating that talent and expertise know no bounds. Her story, popularized by the book and film *Hidden Figures*, continues to inspire women and people of color to pursue careers in mathematics and science.

Vivien Thomas, despite initially being hired as a janitor, showed remarkable skill in surgical research. He developed techniques that significantly improved the outcomes of surgeries for infants with congenital heart defects. Thomas's perseverance in the face of adversity highlights the importance of providing opportunities for talented individuals regardless of their background, thereby enhancing the diversity and richness of the STEM talent pool.

Recognizing these trailblazers is essential in challenging the historical narrative of exclusivity in STEM fields. Their stories illustrate that exceptional contributions can come from anyone, regardless of race, gender, or economic status. By acknowledging and celebrating these diverse figures, we can begin to dismantle the stereotypes that have long persisted in STEM.

Additionally, putting these accomplishments on display can lead to pride and curiosity about STEM in the underrepresented. If kids see themselves in the successful scientists and engineers they learn about, then they are also more likely to believe that similar success is possible for them. This potential for change is an invaluable inspiration that would motivate a broader range of students to enter into and excel in not only STEM education but also a STEM career.

The role of diverse contributions in STEM extends beyond individual achievements—it also includes shifting cultural perspectives within the scientific community. Historically, Western-centric views have dominated scientific discourse, often marginalizing contributions from non-Western cultures. Recognizing the global nature of scientific progress helps to broaden our understanding and appreciation of different methodologies and innovations.

For instance, the mathematical prowess of ancient civilizations such as the Egyptians, Chinese, and Indians laid the groundwork for many modern mathematical concepts. Likewise, Indigenous knowledge systems have made immense contributions to both our body of environmental science and integrative evaluation of sustainable practices. This diversity in cultural perspective benefits the scientific community by providing new approaches and methods addressing much needed global solutions.

Changes in our educational system over the years have also helped more than they hurt, to create a diverse talent pool in STEM. Introducing diverse role models into the curriculum will be important to altering perceptions of who not only can but should belong in STEM. Efforts supporting equal access to STEM education for underrepresented groups such as scholarship, mentorship, and outreach programs will be key in evening the playing field so that individuals from all regions who are qualified can excel.

Organizations and institutions have to simply change on the fly as they come up with plans for supporting diversity in STEM. Having the resources and designing the environment so that underrepresented groups feel welcome and have a sense of belonging can be expected to lead to scientific advancement that is more diverse and cohesive. Diverse thought and experiences bring creativity, which creates more in-depth problem-solving, and original research perspectives.

6.3 Tactics for Drawing More Minorities Into STEM Careers

Evaluating and presenting effective strategies for attracting underrepresented populations to STEM career paths involves understanding and addressing the unique challenges these groups face. To create a diverse STEM talent pool, organizations must implement targeted outreach programs, provide scholarships, facilitate mentorship, and highlight successful case studies.

One of the most important methods that yields strong results is creating outreach targeted specifically to underrepresented high school students. Part of the idea behind these programs is to get students interested in STEM topics early on, by introducing them to different fields and letting them explore through hands-on experiences. Summer camps that focus on STEM subjects may teach students cool projects and experiments. Group mentoring sessions and guest speaker events where STEM leaders present collaborative projects can also spark motivation. The goal is to create a strong STEM push with the foundation of curiosity, passion, and interest in these education disciplines, hopefully turning students into potential young scientists and engineers and preparing them for an inevitable charge toward graduate school.

Examining programs from organizations actively promoting these outreach efforts can provide practical examples and inspiration for new initiatives. One notable case is a program at a Berlin university designed to support female doctoral students and postdocs in science careers. This program features one-on-one mentoring, career development workshops, and networking events. A key strength of the program is matching mentors with mentees who share similar research interests, thereby fostering a mutually beneficial relationship and retaining mentors who transition to new jobs (Women in STEM: Closing the Gender Gap through Effective Mentorship Programs—All Together, 2024).

Time and time again, organizations have discovered the importance of fostering an inclusive sense of community where all students feel accepted. This promotes not just short-term involvement in STEM but also continued interest throughout their lives. By removing the financial, social, and educational hurdles through scholarships, mentorship, and community-building programs, we can diversify our pool of future STEM leaders.

6.4 Corporate and Organizational Structures That Work for DEI Programs

STEM organizations can also create an inclusive workplace by engaging in DEI programs. As such, these programs integrate fairness, equal opportunity, and the concept of inclusion among all employees. To fully appreciate their importance, one has to first delve into how organizations measure the effectiveness of their DEI initiatives, the importance of continuous DEI training, why Employee Resource Groups (ERGs) matter, and practical ways of measuring DEI program effectiveness.

6.5 Assessing Current DEI Efforts

Before any organization embarks on a DEI path, they need to first evaluate their efforts around diversity and pinpoint where improvements can be made, but ensure it is properly aligned with strategic objectives collectively. The assessment would be wide-ranging and examine policies and practices across hiring, promotion, compensation, and employee support initiatives. Surveys and feedback devices can be used to collect input from employees about the experiences and beliefs that are related to DEI within the organization. Companies can also bring in an external third party or consultant to preform independent audits and have no stake on the outcome. This will give one a clear view of where the organization is currently and what will be achievable for the next phase.

6.6 Ongoing DEI Training

Consistently having DEI trainings is another vital strategy for forming that inclusive company culture. One training session does not cut it, DEI has to be continuous and become part of the everyday culture. All employees should participate in these training sessions from leadership to entry-level, so there is a common ground for DEI values and practices. DEI training can cover a wide range of topics such as unconscious bias, cultural competency, and inclusive communication strategies. Training should also be coupled with real-world tools and scenarios to assist employees in navigating more complicated circumstances and having the power to contribute to a positive diverse work environment. Updating the training content to address new issues/trends and challenges in DEI provides a fresh perspective for participants.

ERGs are employee-led affinity groups, based on shared experiences, that support a diverse and inclusive workforce. They are key to establishing belonging and community at work. They create a space for employees to come together and relate to each other due to similarities or troubling life experiences that fit certain characteristics (i.e., race, gender, sexuality, etc.). They can also be useful as a provider of insights for organizational leaders that show the problems and concerns within the workforce. To ensure ERGs are successful, businesses must also support them with necessary resources including funding, time for activities, and executive sponsorship. Recognizing and celebrating the work of ERG members can help drive participation and further the ERG's impact on an organization.

6.7 Assessing the Efficacy of DEI Initiatives

Lastly, a critical piece in the implementation of DEI programs is monitoring their effectiveness to stack up if they are doing what they were made to do. There are numerous strategies companies can use to determine how well their DEI efforts are working. An example, to map outcomes in practice, could mean tracking a diversified composition of staff by race or gender, shift in promotion rates, or aggregate attrition from home-grown talent. Once more, surveys and feedback tools can be used to evaluate employee experience with respect to inclusivity and fairness in the workplace. Organizations can also benchmark their DEI performance against industry standards and best practices to identify areas of strength and opportunities for improvement. Another valuable method is conducting regular reviews of DEI metrics during leadership meetings to maintain accountability and drive continuous progress.

6.8 STEM Diversity Through Outreach Community Programs

Community-based programs play a crucial role in supporting diversity in STEM fields and creating lasting change. These initiatives operate at the grassroots level, often leveraging local resources to engage students from underrepresented communities. By fostering interest and engagement in STEM fields early on, these programs are driving gains toward a more inclusive workforce in the long term.

One prominent example is the Dreamline Pathways program established by A.T. Still University of Health Sciences (ATSU). This community-based initiative introduces K-12 students to health professions through immersive experiences and free educational resources. The ATSU Classroom Kit, which includes tools like stethoscopes and heart rate monitors, offers hands-on learning opportunities that ignite students' curiosity and interest in medical careers (Staff, 2022). These types of programs often serve as the first point of contact for young learners, introducing them to the possibilities within STEM fields.

Community programs can also enhance STEM curricula by partnering with schools. For instance, the Girl Scouts organization collaborates with various educational institutions to integrate STEM activities into their programming. The "Engineer that Girl" event and summer camps focused on LEGO robotics have been particularly effective in sparking interest among young girls. These collaborations not only supplement school education but also provide a practical context for theoretical knowledge, making STEM subjects more relatable and engaging for students (*Bringing STEM Engagement to Disadvantaged Youth through Community-Based Organizations*, 2019).

Outreach is another vital component in promoting STEM diversity. Many community-based organizations operate year-round, offering programs that extend beyond the traditional school calendar. For example, South Carolina's YouthLink runs summer camps and workshops on coding and robotics, ensuring continuous engagement with STEM activities. By providing such opportunities, community organizations keep the momentum going, helping students build a sustained interest in STEM fields (*Bringing STEM Engagement to Disadvantaged Youth Through Community-Based Organizations*, 2019).

Successful campaigns that resonate with potential STEM students and professionals often share common elements: they are accessible, inclusive, and tailored to the needs of the community. The Science Learning Institute (SLI) at Foothill College, established in 2010, exemplifies this approach. SLI operates several innovative programs aimed at advancing equity in STEM, including internships that prioritize underrepresented groups. By addressing financial barriers and offering mentorship, SLI ensures that students not only enter but thrive in STEM disciplines (Staff, 2022).

Evaluating the effectiveness of these programs often involves assessing both immediate and long-term outcomes. Metrics such as increased enrollment in STEM courses, higher retention rates, and improved academic performance are commonly used. However, it's equally important to consider qualitative measures like student feedback and self-reported growth in skills such as problem-solving and leadership.

Almost all the coaches of FIRST LEGO League programs, for example, described substantial progress in their students' STEM learning and interest. Pupils were more receptive to the input of others, more adept at sticking with things when they got tough, and so much better at reinterpreting failure as a natural step to successful learning. These soft skills are invaluable in both academic and professional settings, highlighting the multifaceted benefits of community-based STEM programs (*Bringing STEM Engagement to Disadvantaged Youth Through Community-Based Organizations*, 2019).

6.9 Creating Professionals That Demand DEI-Empowered Employers

One way to foster an inclusive workforce in STEM is to empower underrepresented workers to find and work for organizations that prioritize DEI. This section will offer strategies these professionals can use to better help their job search with research and information, and moving the process further into knowledge-based decision-making.

Initially, professionals must research and evaluate the DEI commitments of prospective employers. Organically they need to start this research by looking at a company's website, checking for visible proof of diversity in the images they use and in their executive leadership team and among their employees. If a company does have DEI as part of their mission statement or purpose, it demonstrates that it has a genuine commitment to it (Saddington, 2021).

Another useful strategy involves evaluating a company's social media presence. Social media platforms can provide substantial insights into an organization's culture and values. One can observe the language and imagery they use, and note whether they actively support causes related to DEI. Companies committed to DEI will engage in conversations about relevant social issues and visibly support key diversity events (Saddington, 2021).

Understanding who a company aligns itself with can also offer clues about its DEI values. One can investigate their partnerships and collaborations, checking if these are with diverse and inclusive organizations. Diverse advisory boards and external engagements signal an authentic commitment to fostering inclusivity.

Reading reviews and comments about the company on various platforms like LinkedIn and Glassdoor can provide additional perspectives. While individual experiences may vary, consistent mentions of DEI-related concerns can be telling. Employee testimonials and comments on social media can highlight how well the company lives up to its DEI promises

Job boards like WORK180 display endorsed employers, ensuring benefits, policies, and perks are clearly visible next to their vacancies (Saddington, 2021). Job seekers can utilize these materials to help them compare DEI pledges among companies and select ones that line up with their own personal values.

Job seekers should check for a company's level of accountability when it comes to the effectiveness of their DEI programs. Applicants are encouraged to ask specific questions about an employer's DEI efforts, or lack thereof; perhaps this will help them identify the key reveals that indicate a company's commitment to diversity. For instance, questions like whether or not they have a chief diversity officer, formal employee training regarding biases in general and anti-racism specifically, and engagement with diverse suppliers can be revealing (Conscious Job Seeking: Assessing Employers' Commitment to DEI, 2020).

One may also want to gauge what a company is doing in terms of gender equity. This type of commitment from businesses to gender balance is increasingly noticeable as more and more organizations are setting targets for the representation of women in leadership (Saddington 2021). While quotas may be viewed with suspicion by some (fittingly so), they are typically a demonstration of an active effort to combat systemic problems.

Professionals need to be able to hold employers accountable with the right support, encouraging companies on positive DEI practices. One practical way to accomplish this is by interviewing a company and asking questions about their DEI path. Relevant questions would be. "What is the company doing to support racial equity?" and "Do they have any affinity groups and are there ways to look up compensation equity analysis?" (Conscious Job Seeking: Assessing Employers' Commitment to DEI, 2020). These responses can shed light on how deep the company goes when it comes to its engagement with DEI principles.

Another tactic is initiating DEI conversations during interviews. List down a series of questions that are more in-depth to the company and their DEI initiatives. This tests an employer's commitment. Inquiries may cover a wide range of topics such as the company's strategies to increase representation at all levels, diversity or unconscious bias training and development opportunities, and how relevant leaders are being held accountable for progressing DEI objectives. (Conscious Job Seeking: Assessing Employers' Commitment to DEI, 2020).

In addition, professionals should not overlook the fact that progress is generally a better indicator than perfection. Frankly, if an organization is truly committed to DEI, it will be more believable in admitting its efforts are a work-in-progress—instead of saying they are perfect at DEI just because they said so, earning itself a reputation for being an engaging and diverse company with an honest accounting of where it is at, down to the challenges.

6.10 Creative Solutions to DEI Initiatives Without Government Funding

Independent DEI Initiatives are a great idea, which is why local assets like businesses and community leaders can be crucial for reaching one's DEI objectives and providing sponsorships, insights, and partnerships. Working with organizations assist groups in building programming that is meaningful. Based on cultural contexts, individual, and community needs. Especially in creating change and building sustaining and inclusive communities. The opportunity is to build regional (state-based) coalitions that can bring like-minded individuals together. These partnerships not only amplify awareness and efficacy of DEI efforts but also foster a sense of ownership and investment among community members.

6.10.1 Utilizing Community Assets

Among the multitude of DEI initiatives of the current day, local community resources are one of the paramount factors in driving change. Without government support, independent DEI efforts can leverage local assets to create meaningful change. One can do this by getting local businesses involved, sponsoring events, and offering in-kind donations. The majority of businesses recognize the importance of promoting diversity and are willing to invest in it. Sponsorship can take many forms, from monetary support to products and services that assist with the implementation of the event. For example, a local restaurant may provide food for an event or a technology company may provide equipment or equipment support. They certainly offset costs, but they also create more participation and awareness of the DEI programs and the businesses involved.

Equally important is the engagement with community leaders to gain trust in communities, because community leaders, trusted voices in their own right, know better than anyone the unique needs and opportunities their communities have. Working directly with these communities to tailor DEI principles to the specific nuances of a given scenario can make them more relevant. This strategy helps create a sense of ownership among community members and assures that community-based interventions are culturally relevant and beneficial. Lastly, these types of partnerships help break down barriers and encourage participation from other demographics, which can only benefit the initiative both in terms of implementation and impact.

Another strategic way to enhance DEI initiatives without a costly bottom line is through organizing volunteer networks. Using volunteers that are passionate and dedicated, can inspire them to get projects done. After developing a dedicated volunteer group, they could potentially reach thousands more people and are able to do much more than would otherwise be possible. This could involve organizing workshops, coordinating events, or providing administrative support. Additionally, the DEI initiative also benefits from having volunteers with varying backgrounds, as this enriches the program with different perspectives and experiences.

Public spaces are less costly and more widely accessible, especially for DEI events. Places like parks, community centers, and libraries work well because they are already outfitted with facilities and community presence for gathering and events. Utilizing these spaces is much more cost-effective as one is not paying high venue rental fees and can instead allocate that money to other critical elements of the initiative. Public venues are also usually situated right in the heart of a city, attracting a much wider audience, and that is crucial for venue inclusivity. Also, hosting events in public spaces inspires action and invites questions from passersby, drawing new supporters and participants into the choir.

Local resources could be highlighted through community partnerships. Joining forces with local organizations, from chambers of commerce to cultural associations, also offers access to a greater pool of resources and a system already in place to tap into local content. Such partnerships can help in reaching out to and bringing in new audiences, and entering collaborations that may be needed for DEI to be sustainable at the grassroots level (Grants for Organizational Diversity, Equity, and Inclusion | NOPI, n.d.).

6.10.2 Corporate Partnerships

It is imperative that strategic partnerships with corporations to advance DEI goals are supported, particularly in the absence of governmental support. When collaborating with corporations, these alliances can lead to visibility, impact, and, ultimately, lasting change. Central to this model are mutually beneficial partnerships that leverage the unique value partnership can bring to both DEI organizations and corporations alike. For the corporation, aligning themselves with DEI goals strengthens their brand, and could help relations with diverse communities, potentially aiding the company's market advantage in the future. The partner organization, in turn, gets access to corporate resources, networks, and knowledge that can significantly advance their DEI efforts. The thinking is that both sides can get what they want, in an increasingly more diverse and inclusive collaboration, and that they can do so through mutually beneficial arrangements.

Corporate partners can deepen their engagement in some of the most impactful ways by offering volunteer hours for their employees. Staff members can offer invaluable talents, perspectives, and energy toward DEI initiatives. Via volunteer programs, employees can give in ways outside of what they do on the job every day, bringing DEI programs those extra hands (and fresh ideas) to initiatives. A more structured DEI volunteerism effort may include details such as how to encourage or make time off available for DEI volunteerism if DEI volunteer opportunities arise or if you can provide recognition of such contributions through enterprise reward programs.

The other key aspect is engaging Corporate Social Responsibility (CSR) programs to fund DEI initiatives. Many corporations have established CSR programs, which are aimed at social issues, so D&I clearly fall under that umbrella. Companies that wanted to pursue particular DEI projects could use some of their existing CSR budgets to underwrite them, so projects that are for the long term can continue to be funded in the absence of a government incentive. Also, initiatives like those that are inclusive to all genders and cultures are good for a company's image. They position the company as a leader in social responsibility, which attracts customers, investors, and prospective employees who value diversity. Transparent reporting about financial spendings and impacts in this context can help build trust and ongoing corporate support.

A different approach to leveraging the shared asset in corporate partnerships is to offer co-branding opportunities. Being co-branded helps the DEI organization as well as the corporate entity, showing that they are working together on solid, evident platforms, events, and marketing campaigns and demonstrating their commitment to diversity. Not only does this partnership broaden exposure of important DEI messaging, but it also showcases the company's commitment to inclusion, an element that can boost customer loyalty. Co-branding can also help send a clear message of partnership and shared values to diverse audiences by collaborating on conferences, co-hosting webinars, or creating joint ad campaigns.

Developing these strategies in parallel means that transparency and clear communication will be key. It demands clarity and alignment in both directions around goals, expected outcomes, and measurements of success. Regular meetings and progress updates can help to ensure that everyone is on the same page and that any new issues are addressed quickly. Formalizing written agreements or memoranda of understanding is another way to build a solid foundation for all of these collaborations, ensuring that there is accountability among the stakeholders as well as a clear understanding of the desired outcomes led by the collaborative effort.

As important as guidelines are for creating and sustaining these types of partnerships, there are dangers in overwhelming stakeholders with unnecessary complexity. Instead, focus on creating practical frameworks that facilitate decision-making and foster adaptability as circumstances evolve. It's also worth noting that successful corporate alliances can set a precedent, inspiring other firms to follow suit. When businesses see how supporting DEI initiatives provides concrete benefits, that can have a ripple effect, encouraging wider participation across the industry. This could also be expanded to include industry forums, trade associations, and networking events that may lend their support and rally corporate peers to the cause at hand.

Lastly, it's important to celebrate wins and share impact stories. Demonstrated success inspires ongoing stewardship and attracts potential new partners. They can also serve as advocacy tools since case studies or testimonials can provide evidence of the success of collaborative DEI efforts. They could shine light on specific, real-world examples of people and communities that directly benefited from those efforts, and that would help drive home the value and impact of corporate–DEI partnerships.

6.10.3 Using Technology and Social Media

Using technology and social media strategically to create and implement DEI could save one a lot of time and money in one's next DEI initiative. One innovation is awareness raising, and reaching wider audiences, through online campaigns. They can also be useful to convey the DEI message, run campaigns, and reach diverse communities. This allows organizations to customize messages that directly resonate with specific audiences to ensure their DEI efforts are on target in the marketplace.

For example, they can focus on gender or racial equity and shape campaigns based on these components of DEI through the use of hashtags to develop a name around their initiative. When organizations partner with influencers and thought leaders within the DEI space, they gain expanded speaking platforms and a reach that engages audiences outside of conventional borders. Integrating interactive features like polls and question-and-answer formats also increases audience engagement, inviting them to participate in the discussion and voice their opinions.

However, beyond awareness campaigns, virtual platforms provide a great funnel to hold meetings and events without the additional logistical costs of holding and traveling to a physical event. The prevalence of Zoom, Microsoft Teams, Google Meet, and related platforms in disseminating information to various stakeholders has transformed the way organizations engage with stakeholders, allowing them the freedom to join from anywhere in the world while avoiding exclusion problems. This virtual model not only reduces costs associated with venue hire and travel but also promotes inclusivity by allowing individuals who may face mobility challenges or other barriers to attend and contribute.

Few organizations have a monopoly on the world's expertise, and digital events help bring speakers and experts from a wide variety of backgrounds together to share their knowledge and expertise. Organizations may organize webinars, panels, or workshops that center around DEI-related topics, allowing space for all to learn and work with one another. Sessions could be recorded and published as educational resources making them potentially more useful over time.

One can also use software tools to follow progress and feedback from participants to ensure that these initiatives are both effective and beneficial. These platforms provide analytics and reporting that allow organizations to measure engagement, evaluate the success of campaigns, and identify areas for improvement. Feedback strategies like surveys and questionnaires offer essential insights into participant's experiences and perceptions, enabling organizations to adjust their strategies accordingly.

Incorporating tools that analyze demographic data can also help organizations ensure that their DEI programs are reaching and benefiting all intended groups, thus reinforcing accountability and transparency in their efforts. Tools like Deel Engage provide dashboards that centralize information, facilitating informed decision-making based on real-time data (30+ Ways to Promote Diversity, Equity, and Inclusion (DEI) in the Workplace, 2024).

Another key component of employing technology in DEI efforts is creating online communities surrounding DEI themes. One of the great things about being a part of these DEI networks is the ability to connect with other DEI professionals—for many DEI practitioners this will be the only DEI professional they know in real life and networking with them can lead to other great opportunities in their DEI journey. Such communities are important networks for ongoing engagement and enable forums for dialogue, sharing of resources, and action. In these spaces, members can discuss challenges, celebrate successes, and co-create solutions, creating a sense of connection and common cause.

This is where innovative ideas and collective action can take root and expand into grassroots initiatives that drive change. Such empowerment can come through mentorships and peer support for underrepresented demographic groups through these networks. Therefore, organizations must be gradual and proactive in building these communities, based on systems for open communication and safe spaces where every voice is valued.

By utilizing technology and social media in DEI initiatives, organizations not only increase their accessibility but also extend their outreach, aligning with modern communication trends and appealing to younger generations that prioritize digital engagement. Adopting these tools allows organizations to develop fluid, agile, and inclusive DEI frameworks that will connect with people from diverse demographic groups. But very importantly, technology, when used wisely, makes all the DEI efforts open, enabling individuals and small organizations with limited finances to pursue a path for meaningful impact without the need for government funding or intervention.

6.10.4 Educational and Training Intervention Programs

Promoting D&I through educational initiatives is crucial for fostering a more equitable society. Such programs can help participants make a difference, informing them so that they are aware of systemic inequalities and empowering them to continue making a positive difference in their positions. This subpoint looks at how to achieve D&I training without state aid by working with public, free initiatives within schools, mentorship programs, certification initiatives, and peer-led training.

One of the most transformative ways to make DEI education a priority is through partnering with educational institutions. With their immense breadth of experience and expertise, schools and universities host tremendous potential for learning and innovation and should be the natural partners for these initiatives. Partnering with academic institutions allows institutions to leverage DEI concepts into existing programs, a step that ensures that the principles of DEI are introduced to students at the earliest opportunity. Organizations can partner with educators to develop custom-tailored workshops that are just specific enough for the audience, while being broad enough to cover the essentials, like unconscious bias, cultural competence, and inclusive communication skills. Not only does this enhance the learning experience for students, but it also prepares them for working in diverse environments. One can also support DEI by offering mentor–intern programs, in which mentees and diverse role models are matched up. These include programs that connect members of underrepresented groups with more experienced, supportive mentors.

Mentoring D&I efforts create certification pathways to ensuring capable, credentialed D&I practitioners. Certification programs also provide structured training that empowers the participant with the tools and techniques to apply DEI in multiple environments. For example, The Diversity Movement has several certifications that are intentional and laser-focused on delivering meaningful, actionable education to make an impact on efforts to create change in workplace culture and leadership (DEI Certifications & DEI Certificate Programs, 2024). These initiatives, also led by Certified Diversity Executives (CDEs), provide a terrific resource for creating diverse, compliance-friendly environments, and remind companies that diversity and inclusion must begin from the top. Through acquiring DEI certifications one can benefit as the certification acts as proof that one is serious about DEI and one can also gain recognition for being an expert, possibly giving one leverage in DEI roles in one's organization.

Certification programs can be designed to meet specific organizational and industry requirements. These could cover topics such as inclusive leadership, employee resource group management, or how DEI plays a role in contributing to change in the company. Instructor-led sessions serve as content for the students in terms of the theoretical frameworks and real-life applications of D&I principles. As a result, these certifications are designed to close the gap between knowledge and action, providing participants with the skills necessary to enact change in their work and in their communities.

Peer-led training is one of the more effective methods for disseminating DEI knowledge across an organization. These sessions, all led by peers, create a collaborative learning environment where those attending can share their own experiences, needs, and ideas. This approach not only encourages open dialogue but also builds a distinctive community that creates more accountability on the part of participants. To work effectively in multicultural settings, one needs diverse perspectives and empathy, which one can gain by learning from one's peers.

In order to bring this approach to the organizations they work for, employees should look for team members who have an interest in D&I to lead these facilitated sessions within their organization. Such facilitators can also be trained with refresher training to brush up on their facilitation skills, ensuring that they know how to structure interactions and lead sets of sessions that the attendees would find engaging and informative. Integrating interactive activities like role-play or case studies can improve the educational process and promote more active engagement. These strategies operate as a coordinated set of solutions; they allow one another to succeed and together shine a light on a tangible path forward for executing DEI education without political pressure dictating the direction. Involvement with community partners can strengthen these activities even further through resources sharing and providing broader access to DEI programs.

6.10.5 Raising Every One's Awareness Through Story and Media

One of the most effective ways to enhance DEI engagement is to produce stories that feature the individuals behind DEI initiatives. Especially at a time when conflicting information can make even basic facts seem dubious, personal narratives place a human face on abstract concepts and offer tangible snapshots of what these efforts look like in practice. If one's stories are true, they foster a level of empathy and understanding that helps audiences connect emotionally to the issues affecting their own lives. In the hands of those who work in an organization, authentic storytelling does indeed break down stereotypes and biases based on the breadth of the experience they have with an organization (The Power of Storytelling in Shaping an Inclusive Workplace, 2024).

Working with filmmakers and writers and developing documentaries and narratives goes to broaden the reach and impact of these personal stories. These ideas surrounding DEI can be easier to digest and more engaging through visual channels. Documentaries are able to address the nuances of D&I, and expose viewers to actual outcomes and triumphs. The stories written are watched in a wider range, so trained writers will take that as a challenge to maintain the interest and readability of the story. While working with the creative talents of filmmakers and writers, stories can be built around DEI issues to amplify the message across the board.

Equally significant are the chances to extend even further, beyond the digital space, employing traditional media channels. While online content is undoubtedly popular, print media, television, and radio are still powerful storytelling platforms, accessed by audiences who may not consume content as frequently online. Some of the best and most effective tactics, like writing feature articles or conducting interviews, anchor DEI stories in ways that invite thoughtful engagement, through a variety of demographic audience lenses. Finally, interviews on talk shows or news segments can shed light on DEI work and reach an audience that often looks to traditional media for information.

Community storytelling events are another dynamic way to platform diverse voices and perspectives. These gatherings create a local event for individuals to share their stories in an encouraging space building community and belonging. Storytelling circles or open-mic nights open dialogue and empower individuals. This enables participants to share their experiences and for listeners to get to know diverse cultures and struggles. Storytelling through community engagement helps to bridge the gap, creating a space for people from all walks of life to meet in the middle and find a common ground and respect for one another hopefully.

But the power of storytelling in promoting DEI only works in a culture where all stories are welcome. Leaders can help this process by modeling openness and vulnerability, by sharing their own stories to set a tone of acceptance and validation (Cecchi-Dimeglio, 2025). It'll help one make sure that nobody is muzzled in one's organization and that storytelling becomes a part of the organization's culture. Leaders need to listen authentically and ensure that different perspectives are embedded into the organizational story.

As storytelling becomes a part of the fabric within DEI strategies, we must be intentional about how we leverage it, both with how we tell stories and how and for whom we gather them. After all, it is the diversity of stories and perspectives that strengthen any initiative that has the audacity to boldly promote change. Stories should share what it means to be human: the challenges and triumphs, and the contributions people make to the social fabric of life.

6.11 Final Insights

In this chapter, we discussed some strategies organizations can put in place to drive diversity and inclusion of STEM talent. Shining a light on historical figures from marginalized backgrounds, we have discovered the extent to which diversity can enhance scientific discovery and upend tired, outdated narratives. Having a variety of viewpoints not only empowers the future but also expands our understanding of scientific accomplishments and processes.

Looking across community programs, outreach initiatives, and DEI commitments that have worked best in companies, it is clear that this work requires ongoing effort and evolution. Initiatives that enable early exposure to STEM and scholarships also contribute toward a conducive learning environment. Together, these initiatives foster an environment where everyone is recognized and supported, generating a more creative and representative STEM community.

Conclusion

In this book, we have journeyed through the myriad factors that contribute to the underrepresentation of marginalized youth in STEM fields. By delving into the historical, economic, and cultural contexts, we've unveiled how deeply rooted societal norms have shaped perceptions of who is deemed competent in science, technology, engineering, and mathematics. The historical narrative of psychological oppression reveals a legacy that still reverberates today, perpetuating cycles of exclusion and disparity. This intricate tapestry of past and present underscores the necessity for an informed and proactive approach to fostering inclusivity in STEM.

As we reflect on these key themes, it becomes clear that understanding the current landscape requires us to recognize these interwoven elements. Historical barriers, socioeconomic challenges, and cultural biases all converge to create an environment where many young people feel alienated from STEM opportunities. By mapping out these connections, we equip ourselves with the knowledge needed to challenge and dismantle these obstacles.

So, now that we've laid the groundwork, let's move to action mode. It turns out that each one of us has a role to play in creating a more inclusive STEM space. If you're an educator, consider incorporating diverse role models and perspectives into your curriculum. Create spaces where every student feels valued and capable. If you're a parent, encourage curiosity and provide resources that spark interest in STEM subjects. Celebrate the achievements of scientists and engineers from varied backgrounds to inspire your children.

Mentorship can be a game changer for those in industry. Helping marginalized students feel confident encourages them to walk through doors they may not have known were open to them. Push for diversity and inclusion policies at your office that lead to fair hiring practices. All of these sound like simple small acts, but when done together, they cause ripples that result in profound change.

Pause for a second and think, what is the ONE thing you can do today to push inclusivity in STEM? This might be getting a conversation going about the need for better science representation or participating in local programs to promote STEM activities for underprivileged young people. Every bit counts toward creating a more just future.

Let us look forward to a world where STEM fields are celebrated as the inspirers of problem-solving minds from every corner and culture. Think of a world in which innovation arises from the unique perspectives and experiences of old, young, rich, poor, and so on A world in which race, gender, or where someone was born does not limit the impact they can make. Imagine a classroom where every child in it, regardless of their background, will have the feeling of belonging and excitement about science experiments and coding projects. It is possible with consistent efforts to be more inclusive; this is not a utopian dream.

In realizing such a vision, community engagement is essential. We need to make way for programs that not only focus on expanding STEM education in underserved areas, but inclusive ones. In doing so, we build a pipeline of broad talent for the challenges ahead. Joint efforts of schools and vocational agencies with private companies are a way to scale these impacts, so that resources and opportunities will be provided to those who need them the most.

But it all does not end here, this is only a start. And for this one requires that one never loses momentum and constantly pushes for more research and advocacy. Continued learning and adaptation is key as we work toward tearing down what now remains of historical oppression. The ongoing conversation is the only way we can continue to progress and garner support for one another instead of competing against each other. It is critical that advocacy remains a constant pillar, advocating policies and practices for inclusion at all levels.

Higher education administrators should support research on the effectiveness of these approaches to diversify STEM fields. This kind of research offers vital information to inform new work. As well, recognizing all the small victories (no matter how few and far between they may be) acts as fuel to the fire and a reminder that there is movement toward progress on a life journey. Because of this, sharing stories of those who have broken down the barriers to be successful in STEM can serve as a great motivation for even more and remind people of why the work must continue to be done on its behalf.

It's a commitment to these principles that allows us to change over the long term. Together, we can leave a lasting, positive legacy of inclusivity for the future. Advocating, educating, and engaging allows the momentum to gain traction instigating an innovative STEM ecosystem.

In conclusion, creating a more diverse STEM is not linear and will result from collaboration from across all silos of society. By acknowledging history, getting involved within our communities, envisioning a more inclusive future and continuous advocacy, we can make the difference! There is much hard work that lies before us, but it can be some of the most rewarding as we move toward a STEM universe in which everyone has a chance to flourish.

So, whether that encompasses mentoring a student, advocating for policy changes, or just spreading awareness, what you are doing now matters. Combining our energies, we can create a world where innovation is unbounded and every capable mind, no matter the origin, makes its mark on the marvels of STEM. We undertake this journey with resolve and aspiration, confident that our shared dedication will lay the foundation for a better, more equitable tomorrow.

References

30+ Ways to Promote Diversity, Equity, and Inclusion (DEI) in the Workplace. (2024). deel. com. https://www.deel.com/blog/promote-diversity-equity-inclusion-workplace/

Access Our Approach, Impact & Efficacy | About Us | PLTW. (n.d.). www.pltw.org. https:// www.pltw.org/about/approach-impact-efficacy

Advancement of STEM Graduate Education: Diversity, Equity, Inclusion and Accessibility. (2024, March 6). NSF - National Science Foundation. https://new.nsf.gov/funding/ initiatives/ige/updates/advancement-stem-graduate-education-diversity

Assari, S. (2016, October 12). *General Self-Efficacy and Mortality in the USA; Racial Differences*. Journal of Racial and Ethnic Health Disparities. https://doi.org/10.1007/s40615-016-0278-0

Barone, R. (2019). *STEM Education Stats for 2019 | Jobs & Careers, Growth, Girls & Degree Statistics*. ID Tech. https://www.idtech.com/blog/stem-education-statistics

The Benefits of a Career in STEM Education – STEAMspiration. (2023, May 20). Steamspirations. com. https://steamspirations.com/the-benefits-of-a-career-in-stem-education/

Hawthorne, G. R. B., Dr Charlene Mickens Dukes, Dr Elizabeth K. (2022, October 17). *Addressing the STEM Workforce Shortage*. www.uschamberfoundation.org. https:// www.uschamberfoundation.org/education/addressing-stem-workforce-shortage

Bringing STEM Engagement to Disadvantaged Youth Through Community-Based Organizations. (2019, January 30). FIRST. https://www.firstinspires.org/stories/bringing-stem-engagement-disadvantaged-youth

Carnevale, A. P., Smith, N., & Melton, M. (2011). *STEM: Science Technology Engineering Mathematics*. Georgetown University Center on Education and the Workforce. https:// eric.ed.gov/?id=ED525297

Cecchi-Dimeglio, P. (2025, January 16). *How Storytelling Enhances Team Decision-Making and Collaboration*. Forbes. https://www.forbes.com/sites/paolacecchi-dimeglio/2025/ 01/16/how-storytelling-enhances-team-decision-making-and-collaboration/

Cobian, K. P., Hurtado, S., Romero, A. L., & Gutzwa, J. A. (2024, January 17). *Enacting Inclusive Science: Culturally Responsive Higher Education Practices in Science, Technology, Engineering, Mathematics, and Medicine (STEMM)*. PLOS One; Public Library of Science. https://doi.org/10.1371/journal.pone.0293953

Conscious Job Seeking: Assessing Employers' Commitment to DEI. (2020). Default. https:// www.naceweb.org/diversity-equity-and-inclusion/best-practices/conscious-job-seeking-assessing-employers-commitment-to-dei/

DEI Certifications & DEI Certificate Programs. (2024, October 28). The Diversity Movement. https://thediversitymovement.com/certifications/

Dost, G. (2024, February 16). *Students' Perspectives on the "STEM Belonging" Concept at A-Level, Undergraduate, and Postgraduate Levels: An Examination of Gender and Ethnicity in Student Descriptions*. International Journal of STEM Education; Springer Science+Business Media. https://doi.org/10.1186/s40594-024-00472-9

Estrada, M., Hernandez, P. R., & Schultz, P. W. (2018). *A longitudinal study of how quality mentorship and research experience integrate underrepresented minorities into STEM careers*. CBE—Life Sciences Education, 17(1), ar9. https://www.lifescied.org/doi/pdf/ 10.1187/cbe.17-04-0066

Easterbrook, M. J., & Hadden, I. R. (2020, October 8). *Tackling Educational Inequalities with Social Psychology: Identities, Contexts, and Interventions*. Social Issues and Policy Review. https://doi.org/10.1111/sipr.12070

Education & Workforce Development – Midwest Big Data Hub. (2024, May 15). Midwestbigda-tahub.org. https://midwestbigdatahub.org/category/education-workforce-development/

Fry, R., Kennedy, B., & Funk, C. (2021, April 1). *STEM Jobs See Uneven Progress in Increasing Gender, Racial and Ethnic Diversity.* Pew Research Center. https://www.pewresearch.org/social-trends/2021/04/01/stem-jobs-see-uneven-progress-in-increasing-gender-racial-and-ethnic-diversity/

Funk, C., & Parker, K. (2018). *Women and Men in STEM Often at Odds Over Workplace Equity.* https://vtechworks.lib.vt.edu/server/api/core/bitstreams/d00f8a3a-a9ac-40df-9e5a-fff4616839de/content

Funk, C., & Parker, K. (2018, January 9). *Racial Diversity and Discrimination in the U.S. STEM Workforce.* Pew Research Center's Social & Demographic Trends Project. https://www.pewresearch.org/social-trends/2018/01/09/blacks-in-stem-jobs-are-especially-concerned-about-diversity-and-discrimination-in-the-workplace/

Grants for Organizational Diversity, Equity, and Inclusion | NOPI. (n.d.). NOPI. https://www.thenopi.org/toolkit/diversity-grants

Haeger, H., Bueno, E. H., & Sedlacek, Q. (2024, March 1). *Participation in Undergraduate Research Reduces Equity Gaps in STEM Graduation Rates.* CBE- Life Sciences Education; American Society for Cell Biology. https://doi.org/10.1187/cbe.22-03-0061

Harnessing the Power of Diversity in STEM Departments. (n.d.). www.herox.com. https://www.herox.com/blog/1080-the-impact-of-diversity-in-innovation

Home | Women Who Tech. (n.d.). womenwhotech.org. https://womenwhotech.org/

How to Retain Diverse Talent in STEM | PA Consulting. (2024). PA Consulting. https://www.paconsulting.com/insights/how-to-retain-diverse-talent-in-stem

Humansmart Editorial Team. (2024). *How Can Organizations Measure the Impact of Diversity and Inclusion Initiatives on Employee Engagement and Retention?* Humansmart.com.mx. https://humansmart.com.mx/en/blogs/blog-how-can-organizations-measure-the-impact-of-diversity-and-inclusion-initiatives-on-employee-engagement-and-retention-56020

The Impact of Online STEM Teaching and Learning During COVID-19 on Underrepresented Students' Self-Efficacy and Motivation | NSTA. (n.d.). www.nsta.org. https://www.nsta.org/journal-college-science-teaching/journal-college-science-teaching-julyaugust-2022/impact-online

The Importance of STEM Education for K-12 Students in Low-Income School Districts. (2023). NMS. https://www.nms.org/Resources/Newsroom/Blog/2023/November/The-Importance-of-STEM-Education-for-K-12-Students.aspx

Jantzer, J., Kirkman, T., & Furniss, K. L. (2021, October 29). *Understanding Differences in Underrepresented Minorities and First-Generation Student Perceptions in the Introductory Biology Classroom.* Journal of Microbiology & Biology Education. https://doi.org/10.1128/jmbe.00176-21

Joecks, J., Pull, K., & Vetter, K. (2013). *Gender Diversity in the Boardroom and Firm Performance: What Exactly Constitutes a "Critical Mass"?* Journal of Business Ethics, 118, 61–72.

Jones, G., Chace, B. C., & Wright, J. (2020, September 22). *Cultural Diversity Drives Innovation: Empowering Teams for Success.* International Journal of Innovation Science; Emerald. https://doi.org/10.1108/ijis-04-2020-0042

K-12 STEM Education for the Future Workforce. (n.d.). Federation of American Scientists. https://fas.org/publication/k-12-stem-for-the-future-workforce/

Li, L. (2022). *Reskilling and Upskilling the Future-Ready Workforce for Industry 4.0 and Beyond.* Information Systems Frontiers; Springer. https://doi.org/10.1007/s10796-022-10308-y

Lorenzo, R., Voigt, N., Tsusaka, M., Krentz, M., & Abouzahr, K. (2018). *How Diverse Leadership Teams Boost Innovation.* Boston Consulting Group, 23(1), 1–8.

Louis Stokes Alliances for Minority Participation (LSAMP). (2015). NSF - National Science Foundation. https://www.nsf.gov/funding/opportunities/lsamp-louis-stokes-alliances-minority-participation/nsf15-594/solicitation

Louis Stokes Alliances for Minority Participation (LSAMP). (2024). NSF - National Science Foundation. https://www.nsf.gov/funding/opportunities/lsamp-louis-stokes-alliances-minority-participation/nsf24-563/solicitation

McGee, V. (2021, July 1). *2021 Guide to Diversity in STEM | ComputerScience.org.* Get an Education the World Needs | ComputerScience.org. https://www.computerscience.org/resources/diversity-inclusion-in-stem/

McKinsey & Company. (2023, December 5). *Diversity Matters Even More: The Case for Holistic Impact | McKinsey.* www.mckinsey.com. https://www.mckinsey.com/featured-insights/diversity-and-inclusion/diversity-matters-even-more-the-case-for-holistic-impact

Melguizo, T., & Wolniak, G. C. (2012). *The Earnings Benefits of Majoring in STEM Fields Among High Achieving Minority Students.* Research in Higher Education, 53(4), 423–452. https://doi.org/10.1007/s11162-011-9238-z

National Academies of Sciences, Engineering, and Medicine; Policy and Global Affairs, Board on Higher Education and Workforce, Committee on Effective Mentoring in STEMM, Dahlberg, M. L., & Byars-Winston, A. (2019, October 30). *Mentoring Underrepresented Students in STEMM: Why Do Identities Matter?* www.ncbi.nlm.nih.gov; National Academies Press (US). https://www.ncbi.nlm.nih.gov/books/NBK552781/

National Science Foundation. (2023). *Diversity and Stem: Women, Minorities, and Persons with Disabilities 2023.* NSF. https://www.nsf.gov/reports/statistics/diversity-stem-women-minorities-persons-disabilities-2023

Northeastern University. (2024, October 4). *Co-op - Northeastern University.* Northeastern University. https://www.northeastern.edu/experiential-learning/co-op/

NSF Scholarships in Science, Technology, Engineering, and Mathematics Program (S-STEM). (2024). NSF - National Science Foundation. https://www.nsf.gov/funding/opportunities/s-stem-nsf-scholarships-science-technology-engineering-mathematics/nsf25-514/solicitation

Okrent, A., & Burke, A. (2021, August 31). *The STEM Labor Force of Today: Scientists, Engineers, and Skilled Technical Workers | NSF - National Science Foundation.* Ncses.nsf.gov. https://ncses.nsf.gov/pubs/nsb20212/participation-of-demographic-groups-in-stem

Ovink, S. M., W. Carson Byrd, Nanney, M., & Wilson, A. (2024, January 10). *"Figuring Out Your Place at a School Like This": Intersectionality and Sense of Belonging Among STEM and Non-STEM College Students.* PLOS One; Public Library of Science. https://doi.org/10.1371/journal.pone.0296389

Pavel. (2023, October 3). *Scholarship and Mentoring: The Key to Recruiting Minority Students to STEM.* Research Features.

The Power of Storytelling in Shaping an Inclusive Workplace. (2024, May 10). Emberin. https://emberin.com/the-power-of-storytelling-in-shaping-an-inclusive-

Project Lead the Way. (2016). PLTW. https://www.pltw.org/

Psico-Smart Editorial Team. (2020). *How Diversity and Inclusion Initiatives Affect Employee Morale.* Psico-Smart.com. https://psico-smart.com/en/blogs/blog-how-diversity-and-inclusion-initiatives-affect-employee-morale-171057

Rincón, B. E., & George-Jackson, C. E. (2016). *Examining Department Climate for Women in Engineering: The Role of STEM Interventions.* Journal of College Student Development, 57(6), 742–747.

Romney, C. A., & Grosovsky, A. J. (2021, October 11). *Mentoring to Enhance Diversity in STEM and STEM-Intensive Health Professions.* International Journal of Radiation Biology. https://doi.org/10.1080/09553002.2021.1988182

Roston, R., Murphy, L., Perera, I., & McGee, E. (2024, September 4). *Lack of Minoritized Inclusion in STEM: Beyond Mentoring to Meaningful Change | Plantae.* Plantae. https://plantae.org/lack-of-minoritized-inclusion-in-stem-beyond-mentoring-to-meaningful-change/

Saddington, J. (2021, May 5). *7 Ways You Can Identify if a Company Is Genuinely Committed to Diversity, Equity, and Inclusion.* WORK180. https://work180.com/en-us/blog/7-ways-you-can-identify-if-a-company-is-genuinely-committed-to-diversity-equity-and-inclusion

Shah, D., Dave, B., Chorawala, M. R., Prajapati, B. G., Singh, S., Elossaily, G. M., Ansari, M. N., & Ali, N. (2024, March 7). *An Insight on Microfluidic Organ-on-a-Chip Models for $PM_{2.5}$-Induced Pulmonary Complications.* ACS Omega; American Chemical Society. https://doi.org/10.1021/acsomega.3c10271

Shortlidge, E. E., Gray, M. J., Estes, S., & Goodwin, E. C. (2024, June 1). *The Value of Support: STEM Intervention Programs Impact Student Persistence and Belonging.* CBE—Life Sciences Education; American Society for Cell Biology. https://doi.org/10.1187/cbe.23-04-0059

Smith, R. (2024, January 9). *Diversity in STEM: What It Is and Why It Is So Important.* https://robertsmith.com/blog/diversity-in-stem/

Staff, I. (2022, August 16). *The 2022 Inspiring Programs in STEM Awards.* Insight into Diversity. https://www.insightintodiversity.com/the-2022-inspiring-programs-in-stem-awards/

Stanislav, E. (2017). *The Physics Book.* Scribd. https://www.scribd.com/document/515193083/The-Physics-Book

The STEM Labor Force: Scientists, Engineers, and Skilled Technical Workers | NSF - National Science Foundation. (n.d.). Ncses.nsf.gov. https://ncses.nsf.gov/pubs/nsb20245/representation-of-demographic-groups-in-stem

Supporting Black/African Americans in STEM - Broadening Participation in STEM | NSF - National Science Foundation. (n.d.). New.nsf.gov. https://new.nsf.gov/funding/initiatives/broadening-participation/supporting-black-americans-stem

Sustainability Directory. (2025, January 23). *Could Inclusive Policies Enhance STEM Diversity?* → *Question.* Sustainability Directory. https://sustainability-directory.com/question/could-inclusive-policies-enhance-stem-diversity/

Taylor, K. (2023, April 18). *10 Top Trends in STEM Education to Follow in 2023.* Kidsparkeducation.org; Kid Spark Education. https://kidsparkeducation.org/blog/10-top-trends-in-stem-education-to-follow-in-2023

Thiem, K. C., & Dasgupta, N. (2022, January 14). *From Precollege to Career: Barriers Facing Historically Marginalized Students and Evidence-Based Solutions.* Social Issues and Policy Review. https://doi.org/10.1111/sipr.12085

Totonchi, D. A., Perez, T., Lee, Y., Robinson, K. A., & Linnenbrink-Garcia, L. (2021, October). *The Role of Stereotype Threat in Ethnically Minoritized Students' Science Motivation: A Four-Year Longitudinal Study of Achievement and Persistence in STEM.* Contemporary Educational Psychology. https://doi.org/10.1016/j.cedpsych.2021.102015

Undergraduate Co-op - Northeastern University College of Engineering. (2024, January 25). Northeastern University College of Engineering. https://coe.northeastern.edu/academics-experiential-learning/co-op-experiential-learning/co-op/undergraduate-co-op/

U.S. Bureau of Labor Statistics. (2024, August 29). *Employment in STEM Occupations.* U.S. Bureau of Labor Statistics. https://www.bls.gov/emp/tables/stem-employment.htm

Index

For Product Safety Concerns and Information please contact our EU
representative GPSR@taylorandfrancis.com
Taylor & Francis Verlag GmbH, Kaufingerstraße 24, 80331 München, Germany